Multiengine Flying

Paul A. Craig

TAB Books
Division of McGraw-Hill, Inc.
New York San Francisco Washington, D.C. Auckland Bogotá
Caracas Lisbon London Madrid Mexico City Milan
Montreal New Delhi San Juan Singapore
Sydney Tokyo Toronto

pb 1 2 3 4 5 6 7 8 9 0 DOH/DOH 9 9 8 7 6 5 4
hc 1 2 3 4 5 6 7 8 9 0 DOH/DOH 9 9 8 7 6 5 4

Library of Congress Cataloging-in-Publication Data

Craig, Paul A.
Multiengine flying / by Paul A. Craig.
p. cm.
Includes index.
ISBN 0-07-013423-5 (pbk) ISBN 0-07-013427-8
1. Twin-engine airplanes—Piloting. I. Title. II. Title: Multi-engine flying.
TL711.T85C73 1994
629.132'5243—dc20 93-46034
CIP

Acquisitions editor: Jeff Worsinger
Production team: Katherine Brown, Director
Patsy D. Harne, Desktop Operator
Lisa M. Mellott, Typesetting
Lorie L. White, Proofreading
Janice Stottlemyer, Computer Illustrator
Elizabeth J. Akers, Indexer
Design team: Jaclyn J. Boone, Designer
Brian Allison, Associate Designer
Cover photograph: Courtesy Cessna Aircraft Co.
Cover copy writer: Cathy Mentzer

PFS
0134235

To Captain Don E. Culp

CAPTAIN CULP WAS AN AIRLINE PIONEER WHO FLEW WITH HOWARD HUGHES, talked over flying matters with Charles Lindbergh, and flew on-call missions for President Roosevelt. Today, Don is an active FAA-designated pilot examiner in Kinston, North Carolina.

Don Culp started firing the powerplants of multiengine airplanes before the FAA invented the multiengine rating. At one time only a "horsepower rating" was issued. If

Charles L. Buchanan

Captain Don Culp in 1993 with an old friend, a DC-3.

a pilot passed a checkride in an airplane with a single 200 horsepower engine, the pilot could act as pilot in command of any airplane with engines that had a total 200 horsepower, plus 50 percent, which equaled a maximum 300 horsepower. This would enable the pilot to fly an airplane with two 150 horsepower engines even though the checkride was in a single-engine airplane. Things have changed.

I credit Don Culp for a vast part of my aviation education. He taught me much about multiengine flying, but more about what it truly means to be a pilot. Pilots are in-flight problem solvers, practical engineers, theoretical scientists, and experimenters. Pilots are also users of common sense, innovators, and respectful of our aviation heritage. Pilots want to know why things work. Pilots are full of life, full of questions, and full of wonder. Don Culp helped me see the important things that mere airplane operators (not pilots) often miss.

I was walking out to an airplane on an extremely hot summer day. The temperature was over 95°F and the heat leaped off the airport ramp and smothered my face. Halfway to the airplane my path crossed with Don Culp as he returned from a checkride flight. We both paused for a moment. In his mid-seventies, when most retired airline captains are taking it easy, he works and teaches and passes on his life-giving lessons. Each of us was sweating from the blistering heat. I was trying to think of the proper complaint regarding the heat, when Culp said, "Well, it's better than blowing snow!" Don Culp also taught me optimism and the idea that flying an airplane is an opportunity to learn, to understand, and to grow.

Thank you, Don.

P.C.
November 1993
Nashville, Tennessee

Contents

Acknowledgments

THIS BOOK BECAME VERY HARD TO WRITE, AND THERE ARE MANY PEOPLE deserving of thanks toward this effort. In the 10 months that I worked on this book, my father-in-law passed away and my family took on the challenge of moving to a new state and university. So the greatest credit goes to my wife Dorothy and the rest of the family, Gabrielle, Ziggy, and the two dogs Aileron and Fowler Flap.

I want to express my deepest felt thanks to Rose Valcarcel for her help these last several months.

Special thanks to John Benton who always had wise advise on this book and so many other things. Thanks, John, for so many great years at LCC!

Eric Kreahemann, a former student, built and wind-tunnel tested the multiengine engine and airfoil section used in the book for data on accelerated slipstream and windmill drag. Thanks again, Eric. Also thanks to Sheila Johnson who built and tested an airfoil for use in a boundary layer experiment.

Thanks to Rick Stowell and PARE for help and advise on this and other projects.

Thanks to Charles L. Buchanan, photographer for the Kinston *Free Press* and instrument pilot. His desire to fly is only surpassed by his talent behind the camera.

Captain Ken Futrell of United Airlines deserves many thanks for help and encouragement from the days when we were starving CFIs together until today.

I have several to thank and appreciate in my new position at Middle Tennessee State University: Don Crowder, Wally Maples, Ron Ferrara, Bill Herrick, Terry Dorris, Billy Cox, Dewey Patton, Steve Gossett, Wade Croissant, and Doug Laue.

Special thanks to Johnny Henley and the gang at ISO Aero Service, Inc., in Kinston, North Carolina, for the opportunity to work and learn.

To my parents Floyd and Anne Craig, I thank you so much. I certainly could not have pulled this one off without you and the Craig Communications, Inc., offices in Nashville.

I have been a chief flight instructor for more than a dozen years now and in that position I am supposed to teach young flight instructors the tricks of their trade. As it turns out I learn as much from them. I have had the privilege of teaching and learning from some of the best flight instructors anywhere: Fred Nauer, Jamie Creel, Wendell Terry, Phil Hietman, Tim Beglau, Dennis Langley, Paul Jones, Latty Bost, Linda J. Smith, Eric Stout, Ken Hamilton, Jamie Smith, and Kevin Bailey. And thanks to my past students and my current group at Middle Tennessee State University.

Introduction

I WENT THROUGH MOST OF MY LIFE WITHOUT PONDERING WHY THE *SPIRIT OF St. Louis* had only one engine. I guess I believed that multiengine airplanes had not been invented yet. But multiengine airplanes had been invented, and they were common. Lindbergh's competitors in the race to Paris were flying multiengine airplanes. Why did Lindbergh insist on an airplane with only a single engine?

Lindbergh had wrestled with a question that many pilots even today do not fully understand. In certain circumstances, a multiengine airplane is not twice as safe as a single-engine airplane; rather it is twice as dangerous.

One of Lindbergh's financial backers in St. Louis suggested that the *Spirit of St. Louis* be designed with three engines instead of just one. One of Lindbergh's rivals was going to attempt the flight in an airplane with three powerplants. Lindbergh responded, "I'm not sure three engines would really add much to safety. There'd be three times the chance for an engine failure, and if one of them stopped over the ocean, you probably could not get back to land on the other two. A multiengine plane is awfully big and heavy. A single-engine plane might even be safer, everything considered."

On another occasion Lindbergh was questioned about his single-engine decision. "Slim," the man asked, "don't you think you ought to have a plane with more than one engine for that kind of flight?" Lindbergh answered, "Suppose one of the engines cuts out halfway across the ocean. I could not get to shore with the other two. Multiengine planes are more complicated; there are more things likely to go wrong with them."

The general public and other pilots perceived that two engines were always better. It would be logical on a flight across the Atlantic that Lindbergh have the best equipment. If you automatically thought "multi" was better, you would have also coached him into a multiengine plane.

What did Lindbergh know that most pilots then and even today did not? He knew that the step to multiengine flying is a serious step. Today we are the beneficiaries of more reliable engines, better aerodynamics, and better equipment than Lindbergh ever dreamed of, yet the potential danger of a multiengine airplane remains a constant.

But the multiengine airplane does have distinct advantages over single-engine airplanes in many areas. The most important advantage is redundancy. Provided that the airplane has enough airspeed, two engines can be a lifesaver. Increases in the thrust-to-weight ratio of engines and better aerodynamics have made it possible for a twin-engine airplane to fly to safety on the remaining good engine. The safety of multiengine redundancy goes beyond the obvious fact that there are two engines. There are also two electrical systems, two vacuum systems, and two fuel systems on most twins. Two engines will often (but not always) produce faster flight speeds. Multiengine airplanes can shrink the aeronautical chart.

The disadvantages fall into two groups: economic considerations and safety considerations.

Economic considerations. Multiengine airplane cost more to buy, operate, and maintain. Having two of just about everything means there are twice as many parts that can fail. Fuel consumption will be at least doubled on a twin, but ground speed will not double.

Safety considerations. Most of this book will deal with the significant risks that exist when one engine fails and the airspeed gets too slow. There is grave danger involved with multiengine airplanes and the *minimum controllable airspeed* (V_{MC}).

There have been attempts to enjoy the benefits of multiengine airplanes (redundancy, speed, and high-speed safety) while eliminating the dangers of V_{MC} (low-speed danger), the most visible of which is the Cessna Skymaster.

The Skymaster beat all the odds. The marketing people inside Cessna did not believe that a twin-engine airplane with one engine in front and one engine in back would be a commercial success, but pilots in the company who knew about the conventional-twin low-speed control problems persevered and the Skymaster became a best-seller.

Cessna sold more than 1,400 Skymasters with retractable landing gear, designated the Cessna 337. There were military versions and even a pressurized Skymaster. Why was it so popular? The design offered the best of both worlds. At the time, it was faster than a single. It could stay in the air and fly to safety with either the front or back engine dead. Best of all, no V_{MC}. When one engine fails, the airplane does not have any additional turning tendency, which is so dangerous with conventional wing-mounted multiengine airplanes.

Cessna originally thought that single-engine pilots would fly the Skymaster without any additional pilot rating, but that idea was dashed when *Flight Standards Service Release No. 467* was published. The FAA required Skymaster pilots to earn a new rating, named *airplane multiengine land—center thrust.*

The Skymaster stopped production in the United States in 1980 and there are still many Skymasters around, but your multiengine flying career will most likely be in conventional twins. This means that you will need to have a healthy appreciation of the pros and cons of multiengine flying.

Flying twins offers some real advantages, but also some real risk. The solution to the risk is knowledge, understanding, proficiency, and hard work. I hope this book gets you started toward those goals.

1
Multiengine aerodynamics

THE BASIC AERODYNAMIC PROPERTIES OF LIFT OPPOSING WEIGHT AND thrust opposing drag still hold true with multiengine airplanes. But multiengine airplanes have more complex thrust vectors, are heavier, and have greater drag than single-engine airplanes. In addition to these basics, the prospective multiengine pilot must learn the difference between performance and controllability.

When flying a single-engine airplane, we typically use V-speeds to reach for the greatest performance from the airplane. The pilot might climb out after takeoff at V_X in order to get over a power line. The pilot knows that by flying the airplane within a close tolerance of this speed she will be getting all that the airplane and its powerplant can offer. Flying a proper V-speed offers an unspoken confidence in safety. The pilot figures that if she flies a certain speed, she is just duplicating what a test pilot has already done. The speed is proven to be not only for best performance but for safety. The speed is reliable.

This line of thinking starts making us believe that V-speeds are only numbers used to provide a certain performance. We assume that a published V-speed is also a speed that provides safety and avoids disaster. Single-engine pilots take V-speeds for granted.

Multiengine pilots must have a deeper understanding of how speeds affect both performance and control. There are V-speeds that a multiengine pilot might fly perfectly and still crash. Most pilots memorize a set of V-speeds just prior to a checkride.

They have a "textbook" definition of each one ready just in case the examiner asks about speed, but they do not understand the speed implications.

STALL SPEEDS

Airplanes must move forward to allow airflow over the airfoils. Anytime the airflow is interrupted or stopped, the airfoil stops producing differential air pressure and lift is no longer produced. The moment this interruption takes place the airfoil stalls. This moment is determined by the angle of attack and not necessarily the speed. So the term "stall speed" is not actually correct. The more correct term is *stall angle*. But most pilots correspond the stall to speed because the airspeed indicator is all that we have to go by. Angle of attack indicators are just not commonly found on light general aviation planes. The airplane, near stall speed, is operating in the region of reverse command (Fig. 1-1).

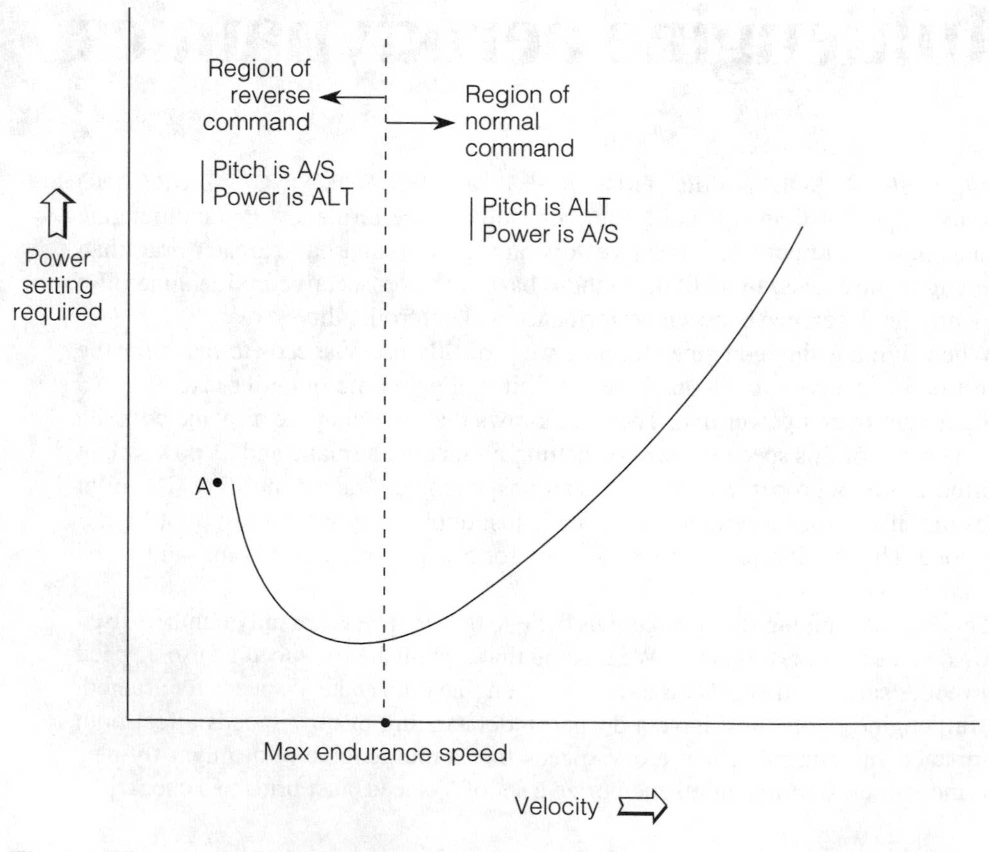

Fig. 1-1. *The power curve.*

In this speed range, the airspeed is controlled by airplane pitch. The airplane's pitch and the airfoil's angle of attack are connected, of course; therefore, speed is controlled by the angle of attack. This works fine until the airfoil's critical angle of attack is reached. When the airfoil stalls, it will do so at a corresponding speed. This speed is marked on the airspeed indicator and is given a V-speed designation depending on the airplane's configuration.

V_{S1}

V_{S1} is the designation for the stall speed (angle) that produces airfoil stall in a "clean" configuration. This means that the wing flaps are up and the landing gear is up. With multiengine airplanes, V_{S1} also means zero engine thrust. When the engines are providing thrust, the slipstream from the engines crosses the airfoils and produces lift. A true V_{S1} stall would be with power off to avoid this effect of the engines producing lift.

V_{S1} is marked by the slow end of the green arc (normal operating range) of the airspeed indicator. Theoretically, the airplane would stall as the airspeed indicator passed below the green arc while experiencing 1G loading. If a loading of greater than 1G exists, an accelerated stall will occur while the indicator still is in the green arc. The airspeed indicator lies to the pilot anytime the airplane has a load factor above 1G. V_{S1} is a control speed. While flying at or faster than V_{S1}, the airplane should remain under control. A situation could arise, especially on a hot day at high altitude, where the V_{S1} speed is being held, but the airplane is in a descent. In this case, the airplane is under control but there is no performance because the airplane is still going down. V_{S1} only guarantees that control surfaces will work and will allow the pilot to keep the sky up and the ground down, but it does not guarantee that the airplane can climb from danger. This V_{S1} idea holds true for both single and multiengine pilots.

V_{SO}

V_{SO} is the stall speed "dirty"—also called stall speed in a landing configuration—because the stall happens with flaps down and landing gear down, just like an approach to landing. Because this speed corresponds to a situation with flaps down, it is slower than V_{S1}. With the flaps down, the airfoil camber is increased and therefore the airflow speed over the camber is increased. This provides more lift and allows this extra lift to be traded for less airspeed.

V_{SO} is marked on the airspeed indicator at the slow end of the white arc (flap operating range) on the airspeed indicator. V_{SO} is the flaps-down control speed. Again, performance is not necessarily provided when flying at V_{SO}, only control is provided. The maneuver routinely performed by student pilots called *minimum controllable airspeed* (MCA) is an experiment in airplane control at slow speeds. The maneuver is not called MPA for minimum performance airspeed.

Many hours have been spent with a student in an airplane with flaps full down, nose high, airspeed between V_{S1} and V_{SO}, full power, and a vertical speed of minus 200 feet per minute. At that moment, V_{SO} will not allow for a climb (performance). In the region of "reverse command," power controls altitude. While at MCA, the high angle of attack and flaps cause a large amount of drag that the engine cannot overcome. At V_{SO}, the airplane is descending but is doing so under control—"behind the power curve." (Position A of Fig. 1-1.)

As a flight instructor, one of my worst nightmares is to have a student pilot on final approach against a stiff wind. The student has full flaps and is low. The student adds ever-increasing amounts of power to remedy the situation, but the underpowered trainer airplane does not have enough power to give. The airplane is under control, but will land in the approach lights and not on the runway.

MULTIENGINE V SPEEDS

The speeds discussed so far apply to single-engine airplanes and multiengine airplanes with all engines operating. This section looks at speeds specific to the problem of multiengine flight. How are these speeds affected when one engine quits on a twin-engine conventional airplane?

V_{XSE}

V_{XSE} (best angle of climb with a single engine) is the speed that will give the pilot the greatest gain in altitude over a certain distance. A pilot forced to use this speed would do so to get out of a real tough situation. Picture a pilot attempting to fly a multiengine airplane off a 1,500-foot airstrip that has 100-foot trees at both ends. One engine quits soon after liftoff. The pilot must clear the trees plus deal with all the problems associated with losing the engine.

The V_{XSE} speed is the best chance (although chances are not too good) to make it out in one piece. V_{XSE} allows, under good atmospheric conditions, the ability to control the airplane and also get the best performance out of the remaining engine. The pilot should only hold V_{XSE} until the obstruction is cleared, and then lower the nose to increase speed.

V_{YSE}

V_{YSE} (best rate of climb with a single engine) is the speed that allows the airplane to achieve the greatest altitude in the shortest elapsed time with just one engine. This speed is on airspeed indicators—a blue line—on multiengine airplanes certificated after 1965. This is often referred to as a "blue radial line." If an engine failed during climbout, V_{YSE} is the best speed to hold if no obstructions are present.

During an obstruction clearance takeoff after one engine failure, the pilot should go to V_{XSE} until the obstructions are cleared and then speed up to V_{YSE} for the remainder of the climb. Single-engine climb performance of light twins is alarmingly poor. The "best performance" might still be a descent, but at either V_{XSE} or V_{YSE}, the airplane should still be fast enough to provide aircraft control.

V_{SSE}

V_{SSE} (safe single-engine speed) is the slowest speed that the manufacturer recommends for practicing engine-out operations. Manufacturers know that pilots in training must shut down an engine occasionally in order to feel and understand how the airplane will fly with only one engine operating; however, manufacturers give the V_{SSE} speed to pilots as a final warning. V_{SSE} is more legal loophole than V speed.

PERFORMANCE AND CONTROL

Understanding airplane performance dominates flight training. We try to understand density altitude and make sense out of eye-crossing charts. We ask important questions: How fast? How far? How much fuel? How much runway? How much weight? Unfortunately, the questions overlook something vital. All the questions assume that the airplane is under control, and that the pilot is in command of the airplane.

Performance is icing on the cake for the multiengine airplane. The actual "cake" is the pilot's ability to manipulate the ailerons, elevator, and rudder to control the airplane. If the airplane ever gets "out of control," those performance charts and their numbers will be of little interest. We begin understanding multiengine aerodynamics by understanding that airplane performance and airplane control are completely different ideas.

A pilot aligns the multiengine airplane with the runway centerline. The voice over the radio says, "You are cleared for takeoff on Runway 5." The pilot adds full power, checks the engine instruments, and the airplane begins to roll. The airplane accelerates to a speed where the pilot rotates the craft into the air. The airplane lifts off and the runway passes behind.

Then, at this delicate moment, one engine coughs and quits. What should be this pilot's first priority: control or performance? Be careful how you answer. It is easy to say that performance is the most important factor because obviously this pilot needs to climb to a safe altitude, and rate of climb is a performance factor. But wait. A climb assumes that the pilot has control of the airplane. The first priority must be airplane control or an airplane climb will not be possible under any performance-limiting situation. Consider the consequences if the pilot does not recognize the control priority.

Multiengine performance is important and is subsequently detailed, but to set priorities straight, airplane control must be discussed first. Some commonsense ideas must be understood. First, two engines are better than one. Second, if one engine quit, adding power to the operating engine would be a necessity. Does common sense hold true in these examples? The answer is a definite maybe. Whether commonsense ideas work in actual practice depends on airplane speed because speed determines control.

The first question: *Are* two engines better than one? If you were flying high over a mountain range at night in a single-engine airplane and the engine quit, you would be in big trouble. If the airplane had two engines and one failed, you would be better off. But "flying high over the mountains" assumes that the airplane is traveling with enough forward speed to be under control.

If the engine quits while a single-engine airplane is slow, for instance during takeoff and initial climb, the airplane is obviously going to come back down. A multiengine airplane might also come back down, even with one engine running. The single-engine airplane should be controllable on the way down (if not stalled), but the multiengine airplane might go out of control. Both pilots are going to crash in this situation, but the single-engine pilot will land wheels down (assuming fixed-gear), and the multiengine pilot lands wheels up. Which is more survivable?

Two engines are very good sometimes and very bad sometimes. How can a pilot determine when two engines are good and when they are bad? Where is the borderline that must be crossed in order to make multiengine airplanes safer than single-engine airplanes? The borderline is a particular speed value called *minimum control speed* or V_{MC}, which is the most important multiengine speed.

The second question: If one engine quits, adding power to the operating engine would be a necessity. Again, if you were flying high above a mountain range and one engine quits, additional power would be required from the operating engine to maintain level flight. When one engine fails on a multiengine airplane, the total airplane drag shoots up and more power is required to overcome this extra drag. Without additional power, the airplane will gradually sink into the mountains.

What about a slower situation, for instance, takeoff and initial climb? If the pilot of a multiengine airplane has one engine fail, there is a range of slow speeds where power from the good engine is a problem. If the pilot pushes the throttles to full power in this situation, the airplane will depart controlled flight and flip, cartwheel, and crash; therefore, two engines are very good sometimes and very bad sometimes. How can a pilot determine when they are good and when they are bad? Where is the borderline between safe and fatal?

Again, the threshold between safe and unsafe is the multiengine minimum control speed (V_{MC}). The FAA has established guidelines to determining this critical speed. Most pilots spend time reading the Federal Aviation Regulation Parts 61 and 91 because those two parts have direct influence on pilots and aircraft operation. Part 23 is unknown to most pilots but is crucial because the rules in Part 23 make an airplane safe and legally airworthy to fly. Part 23 defines the conditions of the minimum control speed (V_{MC}) for multiengine airplanes.

FAR Part 23.149: "V_{MC} is the calibrated airspeed, at which, when the critical engine is suddenly made inoperative, it is possible to recover control of the airplane with that engine inoperative and maintain straight flight either with zero yaw or, at the option of the applicant [for an airworthiness certificate], with an angle of bank of not more than five degrees. The method used to simulate critical engine failure must represent the most critical mode of powerplant failure with respect to controllability expected in service."

The last sentence is key. The phrase "with respect to controllability" clearly means that there is a difference between airplane performance and airplane control that is recognized by regulation.

The regulation then defines the exact condition of V_{MC}: "For reciprocating engine powered airplanes, V_{MC} may not exceed 1.2 V_{S1} (where V_{S1} is determined at the maximum takeoff weight) with:

1. Takeoff power or maximum available power on the engines.
2. The most unfavorable center of gravity.
3. The airplane trimmed for takeoff.
4. The maximum sea level takeoff weight (or some lesser weight necessary to show V_{MC}).
5. Flaps in the takeoff position.
6. Landing gear retracted.
7. Cowl flaps in the normal takeoff position.
8. The propeller of the inoperative engine.
   i. Windmilling.
   ii. In the most probable position for the specific design of the propeller control.
   iii. Feathered, if the airplane has an automatic feathering devise.
9. The airplane airborne and ground effect negligible."

Many unexplained terms in this regulation are essential to understanding the issue. The regulation refers to a "critical engine" and uses the term "windmilling." Phrases such as "most unfavorable center of gravity" and "most critical mode of powerplant failure" appear but are not clearly defined. When it is all boiled down, this regulation is painting a picture of the worst possible multiengine situation that a pilot might get into when it is still possible to recover. If anything gets worse than the situation outlined in this regulation, then there will be no recovery possible and the airplane will depart controlled flight—the plane will crash.

Test ride redline

In order for a multiengine airplane to obtain an airworthiness certificate, a test pilot must take the airplane for a ride. Picture the test pilot up there on a new multiengine airplane's first flight with a full slate of text maneuvers planned. The test crew has loaded the airplane with weights so that the center of gravity is at the aft limit of the CG range and the total airplane weight is at maximum allowable for takeoff.

The pilot climbs to a safe altitude, sets the cowl flaps and wing flaps to the normal takeoff positions (cowl flaps usually open, wing flaps usually up), leaves the landing gear up, and then pulls the throttle of the left engine back to idle. The pilot now banks the airplane 5° to the right—the direction of the good engine—and adds full power to the right engine. Now the pilot allows the airplane's nose to rise and the airspeed to melt away . . . slower . . . slower . . ..

To keep the airplane straight, the pilot is applying more and more right rudder until he can push the rudder no more. The nose of the airplane starts to move to the left, even though the pilot is applying full right rudder. At that very moment, the pilot reaches up with a red grease pencil and marks the position of the airspeed indicator as the airplane rolls over on its back. This is V_{MC}!

Airspeed indicators on multiengine airplanes do have a red "radial" line, although it probably is not actually first marked with a grease pencil (but it is a good joke that Bill Kershner often tells). The red radial line indicates the position of V_{MC} on the indicator, but it is dangerous to assume that you are safe just because the indicator reads faster than the redline because airspeed indicators lie.

The red radial line is placed on the indicator by the "book" value and is based on the FAR Part 23.149 criteria. Pilots can find themselves in situations where the actual V_{MC} is different than the red radial lines indication of V_{MC}. Every multiengine pilot must understand what makes the speed of V_{MC} fluctuate.

Most light twin engine airplanes have one engine on each wing; yet the airplane's center of gravity is in the fuselage. This causes the problem. Neither engine's thrust is pulling through the center of gravity, so yaw forces are produced by engine power. A yaw force makes the entire airplane pivot as if the airplane were stuck on a pin through the center of gravity and twirled side to side. Under ideal conditions, both engines will produce the same thrust and therefore pull the airplane with equal force. But if one engine pulls harder than the other, the airplane will pivot, and the airplane's nose will swing away from the stronger pulling engine.

It is difficult for many pilots to picture the pivot force and understand why the nose moves the way it does while the forces are uneven. To get a commonsense understanding, think of driving your car down a paved road with a sandy shoulder. The wheels and tires providing equal force while on the pavement. The left and right "drive" wheels are turning the same speed to propel the car down the road. Both wheels have equal traction on the road. Both are producing equal thrust. This equal power allows the car to travel in a straight line.

You swerve to miss a pothole and the right wheels drop off the pavement and into the sand. What will happen now? (Fig. 1-2). When the left drive wheel is on the pavement and the right drive wheel is in the sand, they will not produce the same thrust for the total car. This will cause the car to pivot or yaw. Which way will the front of the car sway? In this example, the car will pivot to the right, away from the greater thrust. The left side of the car is trying to "outrun" the right side of the car. The left side moves ahead while the right side lags back and the result is unwanted yaw.

Multiengine airplanes unfortunately do the same thing. Anytime one engine produces more power than the other, the side with the stronger engine pull will try to outrun the weaker side. This will pivot the airplane's nose away from the stronger force. The problem becomes more and more dangerous as the difference between power outputs of the engines gets greater; therefore, the most dangerous situation would exist when one engine was producing full power while the other engine was producing no power. This full power/no power combination would produce a fast and strong pivot force that might be impossible to stop.

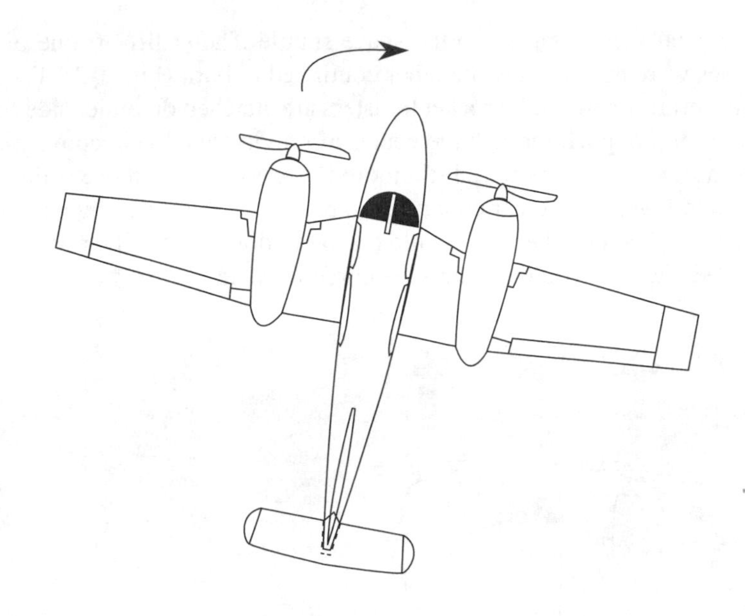

Fig. 1-2. *If the right engine fails, the airplane will yaw to the right just like a car will sway to the right if the right wheels fall off pavement into sand.*

Imagine what would happen to the space shuttle if after lift-off one of the solid rocket engines were to fail while the other continued to burn (Fig. 1-3). The shuttle is a multiengine craft. Its two solid rocket boosters are attached on either side for lift-off. These engines do not push through the center of gravity, just like a conventional light multiengine airplane—except for a little more power! If one of those solid rocket engines were to fail while the other continued to burn, the operating engine would try to continue upward. The failed engine would try to fall downward. The result would be a "yaw cartwheel" which would certainly be unrecoverable.

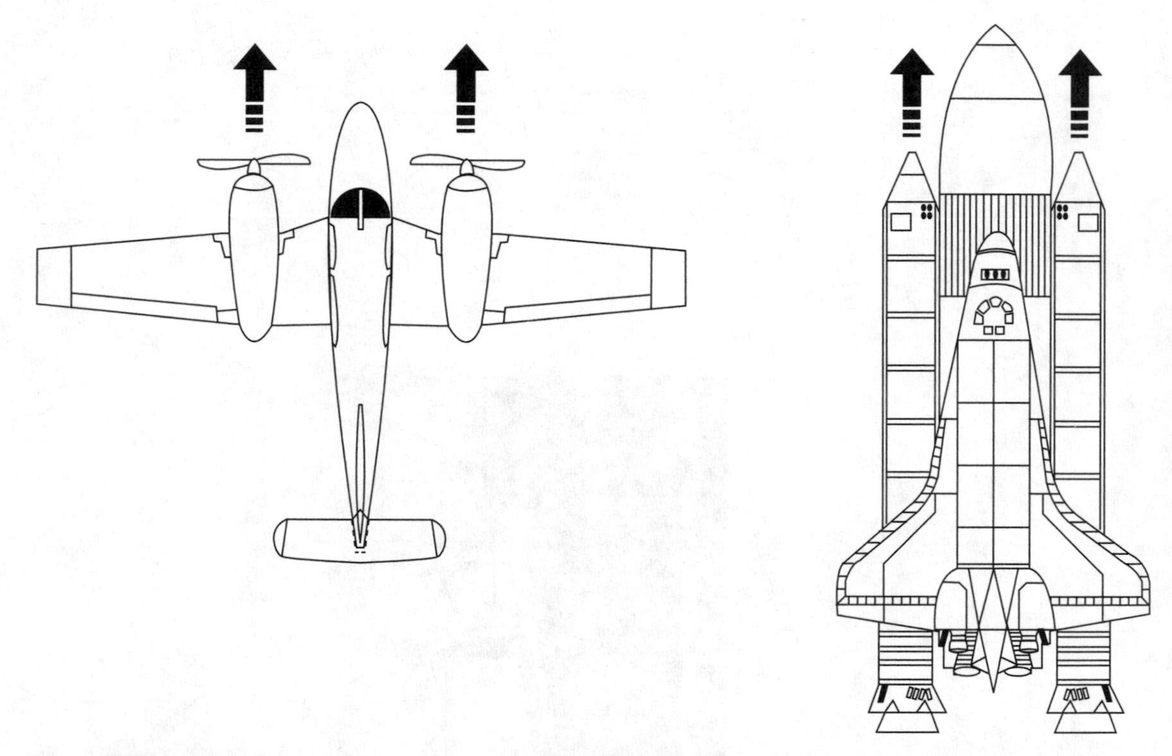

Fig. 1-3. *The twin solid-rocket boosters on the space shuttle do not pull through the spacecraft's center of gravity, much like a conventional twin-engine airplane. If one of the solid-rocket boosters were to fizzle out after liftoff, how much rudder pressure would be required to keep the vehicle on course? What is the space shuttle's* V_{MC}?

If this unwanted yaw occurs in cars, airplanes, and spacecraft, how is the yaw force counteracted? If you were the driver of the car with the left set of wheels on the pavement and the right wheels in the sand, which way should you turn the wheel? The front of the car would be pivoting to the right; to keep the car straight you must turn the wheel to the left in hopes of counteracting the yaw.

Your only hope of keeping the car straight would be if the steering wheel's left turning force is at least equal to the car's right turning force. If both forces are equal, they will in effect cancel each other out, and the car will continue straight ahead. But if the forces are not equal, the car will continue to pivot, maybe to a point where control is lost.

What provides the counteracting force in an airplane? Airplanes do not have steering wheels that provide any help to prevent pivot. Airplanes move in three axis: pitch, yaw, and roll. This pivot takes place in the yaw axis (Fig. 1-4), which is controlled by the rudder. Of course the airplane control surfaces require airflow to function properly. All pilots test the control surfaces during a pretakeoff check to ensure that they are moving in the proper direction; but the airplane does not pitch, or roll, or yaw, because the airplane is standing still with minimal airflow. Airspeed makes the control surfaces effective.

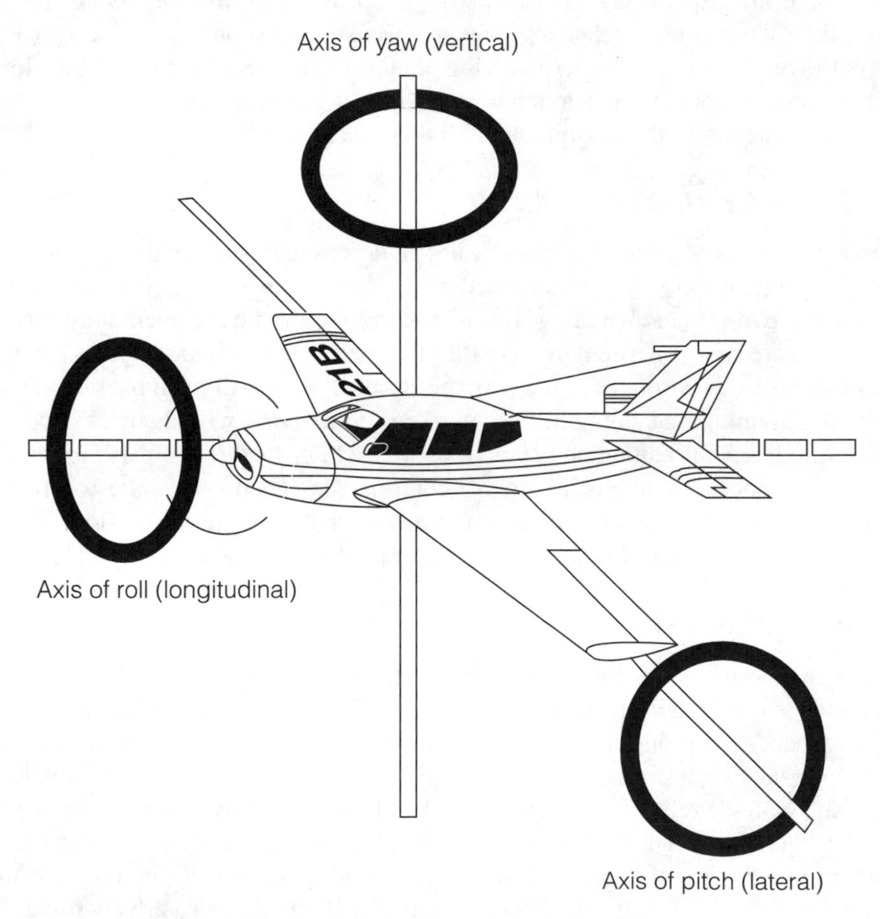

Fig. 1-4. *The battle for control is fought in the yaw axis.*

My first instructor wanted me to feel how "sluggish" the flight controls were during slow flight and stalls. This sloppy feel of the controls was due to their lack of effectiveness, which in turn was due to the reduced airflow at slower speeds. So the pilot's ability to make the airplane yaw by using the rudder is a function of just how fast the airplane is traveling through the air. If the pilot needs the yaw force of the rudder to counteract an uneven thrust from the engines, the pilot's ability to overcome this force depends on airspeed.

A car driver uses the steering wheel to overcome unwanted yaw. An airplane pilot uses rudder to overcome unwanted yaw, but the rudder does not always give the pilot what is needed.

Many forces are at work when uneven thrust exists. The "pulling ahead" force and the "lagging back" force produce the yaw in one direction. Airflow past and around the rudder can produce yaw in the opposite direction. If all these forces are equal and cancel out, then the airplane can continue straight ahead under control. As airflow gets slower, the rudder counterbalancing yaw gets weaker and soon the engine yaw overpowers the rudder yaw. This overpowering point is the airspeed that is just too slow to make the rudder work strong enough to counteract the uneven engine yaw. This exact speed now depends on the magnitude of all these factors.

FACTORS AFFECTING V_{MC}

The exact speed of V_{MC} fluctuates and is not always as indicated by the red radial line on the airspeed indicator. The exact speed of V_{MC} is actually a battle between opposing forces. Like two armies marching toward each other, the forces eventually clash. If both armies are of equal strength the battle line will become stationary. But if either army emerges as the stronger of the two, the stronger army will push back the weaker opponent. Anything that strengthens or weakens either army will change the position of the battle line. Our battle line is drawn by V_{MC}. On one side of V_{MC} is aircraft control and the opposition side is no aircraft control. The "good" side relies on fast airspeed and the weakness of the enemy's force. The "bad" side relies on slow airspeed and its own yaw strength. Certain forces can strengthen or weaken the "bad" side.

Power produced

The greatest pivot/yaw force would occur when one engine is operating at maximum power and the other engine has zero power. They would only occur on takeoff, outside of a training maneuver. If one engine failed and the other continued to operate at maximum power upon takeoff or during initial climb, the yaw would be the greatest. Couple that strong yaw with the relative slow speed of takeoff and the ability to counteract the yaw would be at its lowest. This is why takeoff in a multiengine airplane can be so dangerous.

More power from the engines usually is a good thing. But if one engine should fail, any "extra" power from the operating engine will produce greater yaw force. This additional yaw force can only be counteracted with additional rudder yaw force, but at slow speeds this "additional" force might not be available or even possible.

So, in an engine out situation, extra power from the good engine is bad. Anything that allows the power produced from the engine to become greater strengthens the "bad" side of V_{MC}. This requires a faster speed to overcome the extra yaw. Anytime more speed is required for control, the "bad" side is winning the battle.

Several factors can cause the engine to produce more power, but the most influential is density altitude. Engines breath air. Reciprocating engines draw air inside, mix it with fuel, squeeze it with a piston in a cylinder, and burn it with a spark plug. Turbocharged engines do the same thing, except they presqueeze the air before sending it to be burned. Turbine engines do the same thing. They draw in air, squeeze it with a rotor, mix it with jet fuel, and burn it in the combustion chambers. Even rocket engines in space breathe air; the spacecraft carries air as liquid oxygen.

The engines draw in air, which is a mixture of many gases. The gas that we are particularly interested in is oxygen because oxygen burns like a fuel. The other gases that get pulled into the engine are used as well. They get hot and expand, which helps push a piston down, but oxygen causes the fuel to give up its energy.

Because the air is a partner to fuel in combustion, it makes sense that when there is more air, there is more combustion and vice versa. Why do hot coals glow red when blown upon? Because more fuel (air) is added. Why does a candle eventually flicker and go out when left burning in a jar? Because the fuel (air) is all used up. In an engine, when the content of air goes down, the combustion goes down and less power is provided from the engine.

Gravity holds the atmosphere tight to the Earth. At sea level, the atmosphere's gas molecules are crushed together under the weight of the atmosphere. Humans have a hard time understanding this "weight" because we simply do not feel it. The pressure of this weight is inside and outside our body. We do not feel the weight because everything cancels out.

Low in the atmosphere, near sea level, the air flows like a river into engines. But high in the atmosphere, above the majority of air molecules, engines gasp for air. To the engine, climbing up into the atmosphere is like slowly turning down the fuel supply. The engine gradually loses power until it can climb no more. Power loss with a gain in altitude restricts aircraft operations.

Many pilots have been forced to go around a mountain range instead of going over it. The limiting factor is the ability of the airplane to get well above a ridge and its turbulence. Density altitude considerations are usually a no-win situation. Anytime density altitude is considered, it usually means aircraft performance is in question. The outcome is a loss in performance. Density altitude is a paradox for multiengine aerodynamics: disadvantageous to performance, advantageous to V_{MC}.

Anytime the power of an engine is reduced, the ability of that engine to produce yaw is also reduced. Less yaw from the engine relaxes the need for the rudder to counteract yaw forces. The airplane requires less airflow across the rudder to keep the nose straight; therefore, V_{MC} gets slower.

A cold day at sea level will allow the engine to produce its maximum power, which is bad for V_{MC}. A hot day in the mountains will deprive the engine of its power, which is good for V_{MC}. But remember, a good situation for V_{MC} might not necessarily be good for performance. Here again, control of the airplane and performance of the airplane are two completely different things.

Power can also be reduced by simply pulling the throttle back. Manually reducing power on the good engine improves airplane control by reducing V_{MC}. The problem is that the airplane might not climb or even stay level with a reduced power setting on the good engine. Here is the crux of the matter. Low to the ground and with an airspeed at or slower than V_{MC}, a pilot does not have any good choices. The pilot must reduce power on the good engine for a slower V_{MC} and stay under control.

This reduced power setting is not strong enough to pull the airplane to a safe altitude. The pilot needs to understand that no matter what he does, the airplane is going to crash. The only choice is to crash right side up or upside down. Instinctively, pilots will push the throttle forward in a futile attempt to climb away. But increasing the power on the good engine produces extra yaw that cannot be overcome by rudder at this slow forward speed. As the power comes up, V_{MC} comes with it and the airplane begins to roll and yaw out of control.

The only hope of a noncrash recovery is altitude. The pilot could reduce the power on the good engine and therefore reduce V_{MC}. Then, using the altitude below the airplane, lower the nose and trade altitude for airspeed. As the airspeed gets faster, the pilot can afford to add more power because the additional speed will allow the rudder to overcome the additional yaw. If enough space is remaining between the ground and the airplane, the pilot can continue walking this speed/control tightrope until the available performance is achieved.

This entire idea is foreign to our original single-engine training. I was told by my first flight instructor to go-around anytime that I was not completely comfortable with a landing approach. Go-around means full power. So we associate "get out of trouble" with "full power." But in certain multiengine circumstances, for instance, low to the ground and slower than V_{MC}, full power is fatal.

Multiengine pilots must do everything possible to avoid that situation. But if such a situation does arise while one engine has failed, the pilot must have the in-depth understanding that will allow a split second decision to pull back on the power and then accelerate, or if no altitude is available, to accept a forced landing. It is simply against the pilot's nature to sit quietly, holding throttles back, and watch the ground come up. Pilots cannot allow themselves to get painted into this corner.

Uneven drag

Another misunderstanding: When one engine fails, 50 percent of the total aircraft power disappears. Even though the power is being reduced by half, the aircraft performance might be reduced as much as 80 percent. The difference is due to the increased drag that can form on the airplane with one engine out (Fig. 1-5). Recall the concept that the good-engine side causes yaw by attempting to "outrun" the bad-engine side. Now consider that drag will pull the bad-engine side rearward, while the good-engine side pulls forward.

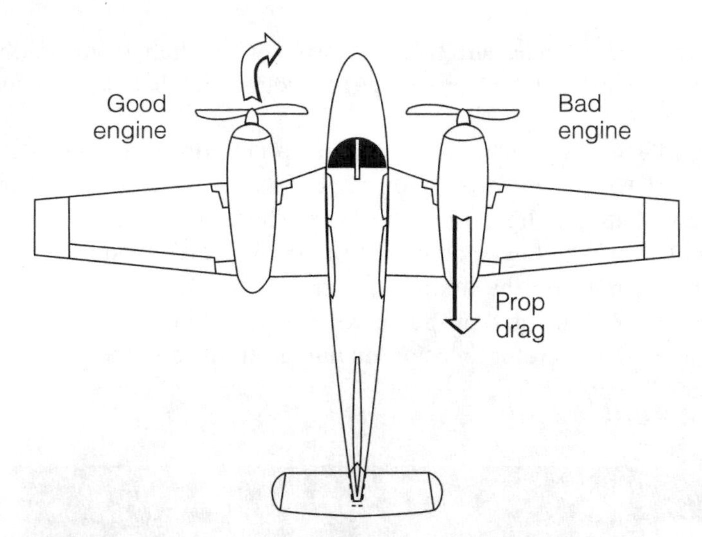

Fig. 1-5. *When half the power is lost due to an engine failure, as much as an 80 percent loss in performance can result due to the uneven drag.*

The pilot is in between these two forces, sitting on the pivot point. Left alone, these two forces will "flat spin" the airplane. The pilot must hope that there is enough rudder effectiveness to overcome this twirl. The airplane will require considerable forward speed and corresponding airflow over the rudder to stop this dangerous yaw from taking place. The only way out is to reduce the forward force on the good engine's wing (reduce power) and reduce the rearward force on the bad engine's wing (reduce drag). The propeller that is not producing thrust has become a liability by causing drag.

When one engine quits, the engine instruments do not immediately tell the pilot which engine failed. The airflow will continue to turn the propeller like a pinwheel. This is why you cannot determine the failed engine just by looking. The propellers will look as though they are both running just fine. The propeller of the dead engine is turning, which causes the tachometer to read as it always does. There might be a slight reduction in RPM because relative wind, not combustion, is causing the crankshaft to turn.

The manifold pressure gauge is also no help. The wind that drives the propeller is also driving the pistons to rise and fall in the cylinders; the valves are opening and closing with the proper rhythmic timing. Every intake stroke draws air in past the manifold pressure port that reads the airflow just as any normal situation.

The propellers look normal, and the tachometers and manifold pressure gauges look normal, but the nose is yawing and great amounts of rudder are required to keep the nose straight and the airplane under control.

Propeller drag must be reduced. The propeller can do one of three things:

- Left alone to fan in the breeze.
- Blades stopped.
- Blades turned into the wind.

The condition where the blades turn by the pressure of the relative wind alone is called *windmilling*. The condition where the blades turn edge-on to the relative wind and stop rotating is called *feathering*.

A test airfoil was designed to prove the drag production of the propeller in three conditions (Fig. 1-6). The airfoil section represents the wing section that holds the engine of a conventional light twin airplane. The airfoil/engine combination is shown in Fig. 1-6 in the slow-speed wind tunnel where the tests were completed. The propeller was powered by a small electric motor. In the series of tests conducted for drag, the engine and propeller were installed but the electric motor was turned off. The engine therefore was producing no thrust and the propeller became a drag producer.

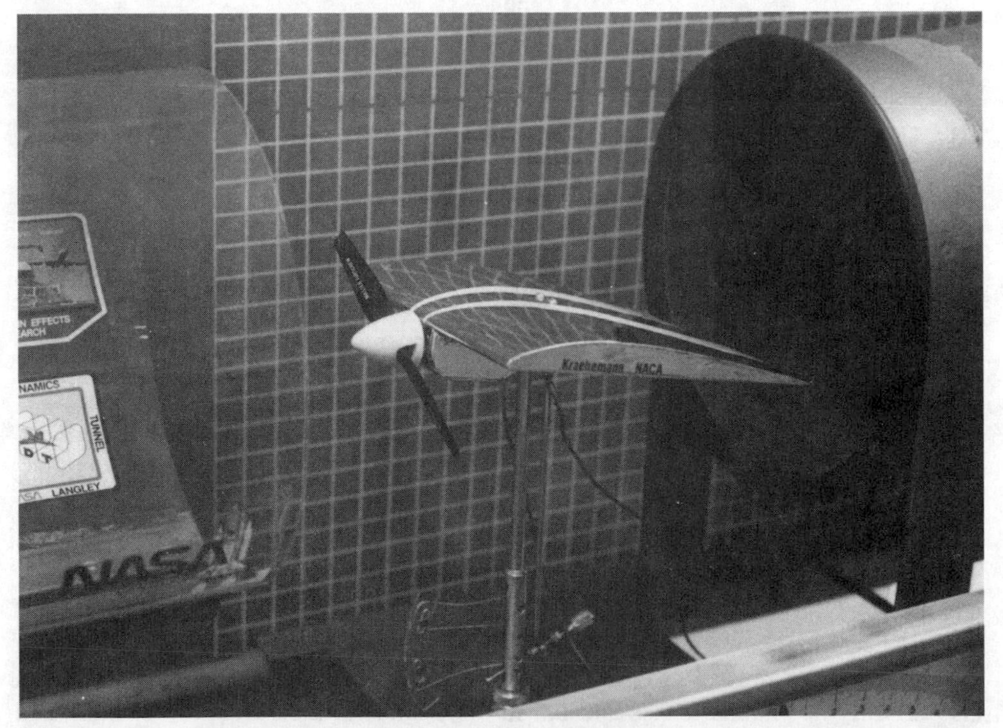

Fig. 1-6. *This airfoil/engine combination represents the wing section of a conventional multiengine airplane where the engine is mounted. The test airfoil/engine was placed in a wind tunnel to test propeller drag in three configurations: windmilling, propeller stopped, and propeller feathered.*

In the first test, the limp propeller was allowed to windmill in the tunnel. The tunnel's airflow turned the propeller just as relative wind turns a dead engine's propeller in flight. Figure 1-7 displays the test results of the power-off, propeller-windmilling test.

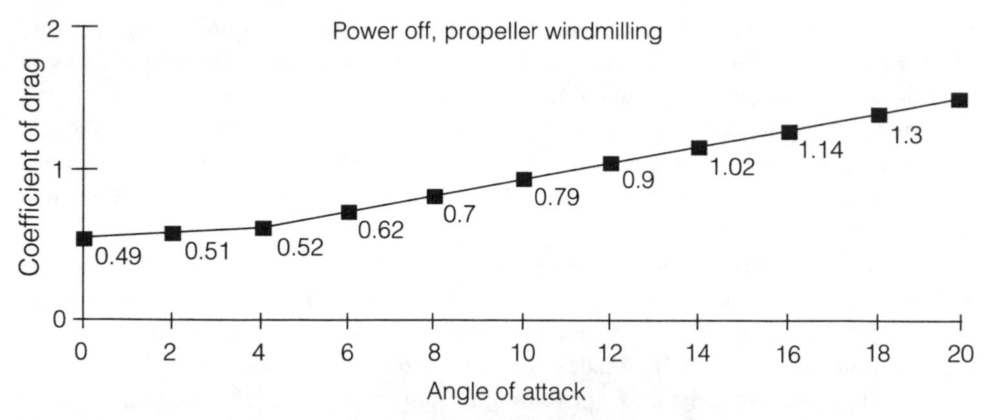

Fig. 1-7. *Coefficients of drag produced by a windmilling propeller.*

In the second test, the propeller was stopped and prevented from windmilling in the airflow (Fig. 1-8). Comparing the two graphs tells the story of propeller drag. Examine the coefficient of drag at 8° angle of attach on each graph as one example. When the propeller was windmilling, the C_D was 0.7. With the propeller stopped, the drag was reduced to a C_D of 0.56. That is a 20 percent reduction in drag just because the propeller was prevented from windmilling. A turning propeller is essentially a disk in the wind (the diameter of the disk is the diameter of the propeller arc), not simply two or three blades in the wind.

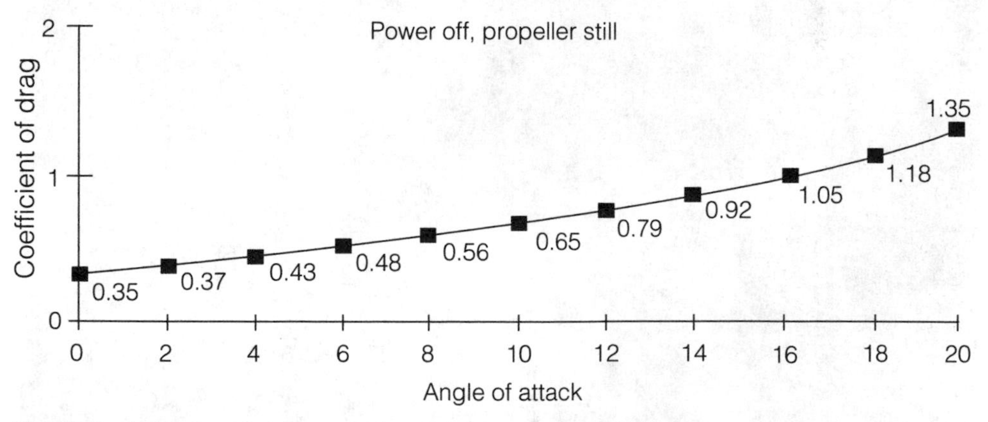

Fig. 1-8. *Coefficients of drag produced by a stopped propeller.*

(This principle applies to single-engine airplanes as well. The pilot can get a much better glide distance—20 percent better in this example—by stopping the propeller of a single-engine airplane. The pilot would have to raise the nose and reduce the airspeed. At a certain speed, the drag from within the engine would be greater than the airflow and the propeller would stop. For the remainder of the glide, the pilot would have more choices. Be careful because raising the nose to reduce the airspeed has its hazards and should be done with plenty of altitude.)

For multiengine airplanes with one engine failed, this reduction in drag would cause the airspeed of V_{MC} to get slower. Less drag on the dead engine side would mean less rudder to overcome the drag. So, any drag reduction caused by the propeller will be good for V_{MC}.

Propellers on multiengine airplanes have one more drag reducing feature: feathering. When the propeller blades are stopped, the airflow still strikes the blade on the face. The propeller blade is designed to be streamlined when rotating. Without rotation, the relative wind hits the blades head-on. A propeller feathering system allows the propeller blades to twist into a position that is more streamlined. The relative wind then hits the blades edge-on and drag is further reduced.

The last drag test using the sample airfoil was conducted using a feathered propeller (Fig. 1-9). This photo shows the twist on the propeller blades. As before, the tunnel's airflow was allowed to pass by the airfoil in the feathered position. Figure 1-10 shows the feathered test results. Comparing as before the 8° angle of attack setting, the coefficient of drag is 0.43. This is a 38.6 percent reduction in drag compared to the windmilling condition.

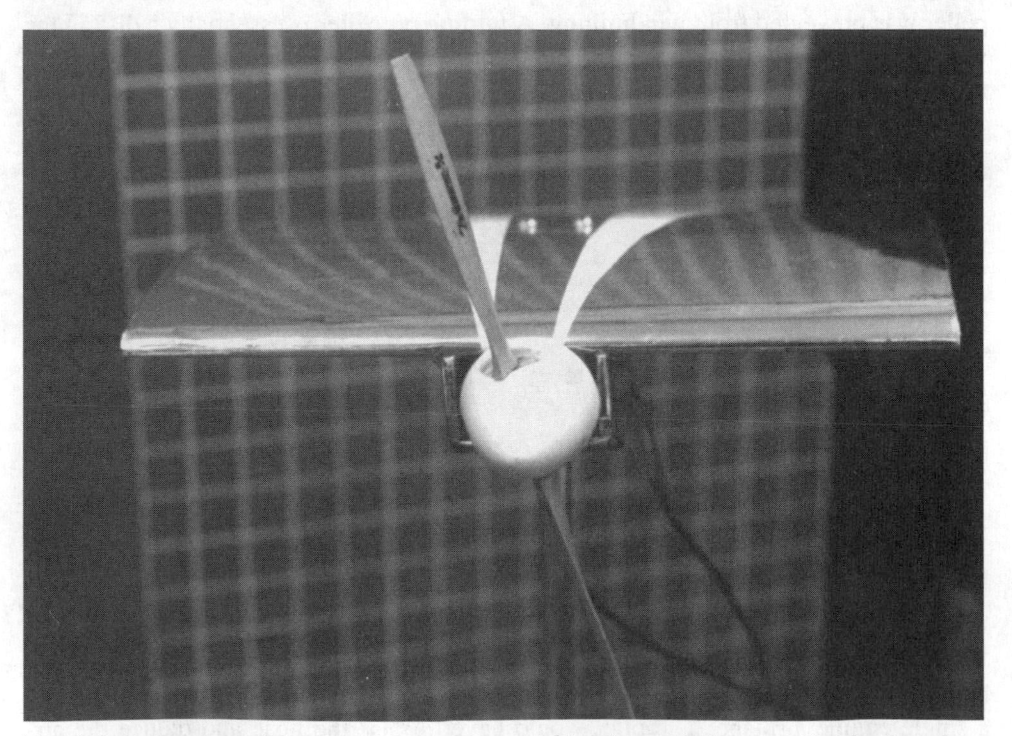

Fig. 1-9. *Front view of the test airfoil/engine combination with the propeller feathered.*

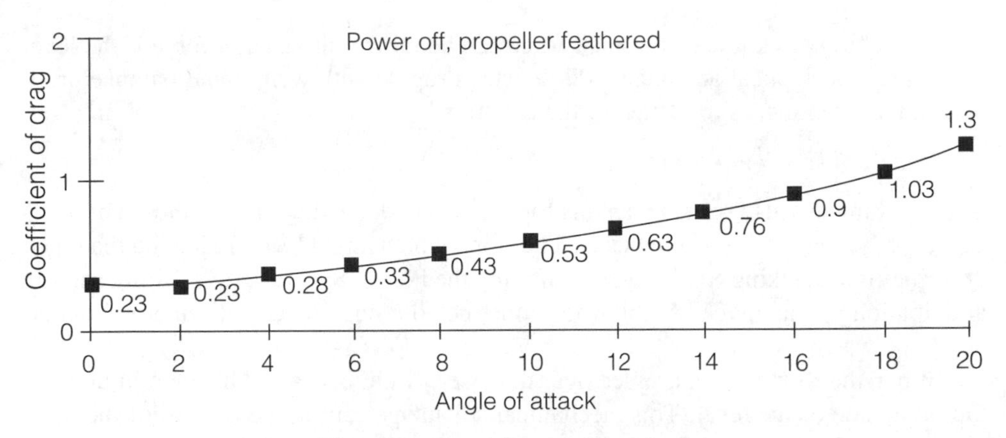

Fig. 1-10. *Coefficients of drag produced by a feathered propeller.*

These results are from an independent test of a generic airfoil and engine and do not necessarily represent the drag savings of any particular airplane. The results do illustrate the science involved with any multiengine airplane.

If a multiengine pilot experiences a low altitude/slow airspeed failure of one engine, the sooner that the dead engine's propeller is feathered, the safer the situation will be. A feathered propeller produces less drag, which requires less rudder effectiveness and airspeed. A feathered propeller is good because it causes V_{MC} to become slower. A windmilling propeller forces the dead-engine side rearward, which requires a strong rudder to pull the dead-engine side forward. Windmilling increases V_{MC}.

THE LAW OF THE LEVER

To this point, the forces discussed were not amplified. The forces shown in Fig. 1-5 were relatively simple. One force pulls forward; the other pulls rearward. But forces present in a multiengine airplane become more and more complex the closer you look. Not only is the presence of a force important, but the location of the force is important and might have a great impact on the eventual magnitude of the force. This is a very old idea applied to the multiengine airplane.

The idea is so old it goes back to biblical times. Pilots do not own the science of weight and balance; a book on the topic was published as early as 250 B.C. entitled *Centers of Gravity of Planes* by Archimedes. I do not think the term planes in the title meant airplanes. Archimedes was a famous Greek scientist and mathematician. He understood something before the birth of Christ that multiengine pilots need to know today. He wrote the law of the lever: "Magnitudes balance at distances from a fulcrum in inverse ratio to their weights."

Loosely translated, this means that two kids on a teeter-totter will only balance if the kids weigh the same. It also means that a lighter weight could balance a heavier weight as long and the lighter weight has a longer arm (distance) from the balance point. Archimedes knew that a small force can be turned into a large force if the lever is used for mechanical advantage. Pilots who struggle with weight and balance problems have Archimedes to blame for the equation

Weight × arm = moment

(Archimedes also became famous for his help in defending his hometown of Syracuse, Sicily, against Roman invaders. The story, which might be more legend than fact, describes the attacking Roman Army entering the Port of Syracuse on a sailing ship. In anticipation of the attack, Archimedes constructed a huge lever with one end underneath the water's surface.

When the Roman ship crossed over the lever, Archimedes and his men jumped on the other side of the lever. This mechanical advantage caused the far end of the lever to rise from the water and flip the ship. Archimedes died in 211 B.C. when the Romans attacked Syracuse on land instead of water, but the law of the lever survived.)

When dealing with uneven forces, such as uneven thrust or drag, the position of the force determines the magnitude of the force. V_{MC} determines the speed required to overcome the force, so V_{MC} and the law of the lever must be understood together.

THE CRITICAL ENGINE

To understand why one engine might be considered more critical than another engine, basic propeller theory must be understood. Recall the terms *P-factor* or *asymmetrical thrust*. Both terms refer to an inevitable turning force that results when a propeller turns in an inclined geometric plane. Most pilots have heard about the force, but few truly understand it. We know that during a full-power climb in a single-engine the airplane tends to turn left and that right rudder cures the problem. But why does the airplane turn? What does turning have to do with V_{MC}? What does all that have to do with Archimedes?

When the airplane is flying straight and level, the thrust from the propeller is centered. The descending propeller blade cuts through the air at a particular angle, and this produces forward thrust. At the same time, the ascending blade makes the same size cut into the air and therefore produces the same amount of forward thrust. When thrust from both sides of the propeller disk is equal, the thrust pulls as if through the spinner.

Things change when the airplane is in a climb with the nose pitched up and the entire airplane tilted back (Fig. 1-11). This tilt increases the descending blades' cut into the air, simultaneously reducing the ascending blades' cut into the air. The descending blade will produce more thrust because the blade is taking a bigger bite of the air; the ascending blade will produce less thrust because it is taking a smaller bite of the air.

The propeller's *thrust pattern* has shifted. Initially, the effective thrust came from the propeller's spinner. Now, with the nose inclined, the focal point of the thrust shifts toward the descending blade. The center of thrust moves from the center of the propeller to somewhere closer to the descending blade. A single-engine airplane's propeller turns clockwise as seen from the pilot seat. This means that the descending blade is on the right side. When the center of thrust shifts to the right side, the airplane will turn left be-

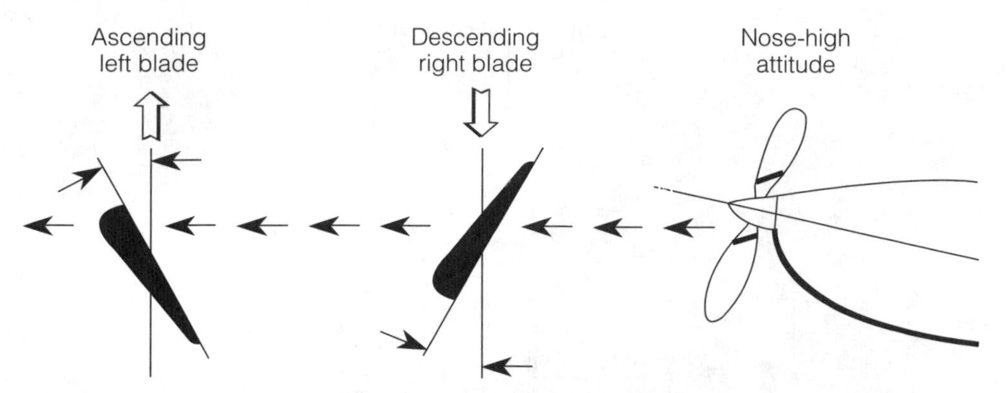

Fig. 1-11. *P-factor is caused when the descending right propeller blade takes a larger bite of the air than the ascending left blade.*

cause the right side of the blade has more thrust and tends to outrun the left side. This causes the entire airplane to turn left and pilots stop this turn with right rudder.

Apply this to a multiengine airplane climb. Both descending blades will be taking a bigger bite of the air; both thrust patterns will be shifted to the right. Figure 1-12 shows the thrust from both sides of the turning propellers. The left-side thrust arrows are shorter than the right-side thrust arrows, indicating the greater force on the right. The airplane's right side will try to fly faster than the left side. But because both sides are connected by the rest of the airplane, the result is a left turning tendency. This situation is normal and presents no real problem. All a pilot must do to correct the situation is apply right rudder.

What would happen if one of the two engines quit while this uneven thrust was taking place? The airplane would yaw in the direction of the dead engine. This fact has already been established, but the *critical engine* concept considers the degree of yaw force. In this situation shown in the diagram, a failed left engine would produce a greater degree of yaw force than a failed right engine. The reason the left engine is the "critical" engine takes us back to the lever and Archimedes.

The pivot point of the airplane is the center of gravity. Usually when we think about center of gravity, we are attempting to properly balance the airplane. Using standard weight and balance theory in an elementary sense, we place weights of different magnitudes along the line from nose to tail. Each weight is a force pointing down to the center of the Earth that must be offset in order to balance the fuselage. Considering V_{MC}, we must view the forces from above and look along a line from wingtip to wingtip. The forces as described are thrust vectors of different magnitudes that must be offset not by weight but by rudder force.

The farther a force is from the center of gravity, the more leverage it will have. The force is simply magnified by its position relative to CG. Looking at the diagram one more time, which propeller force is farthest from the CG? The force coming from the right engine's descending propeller blade is farthest. This means that in a situation

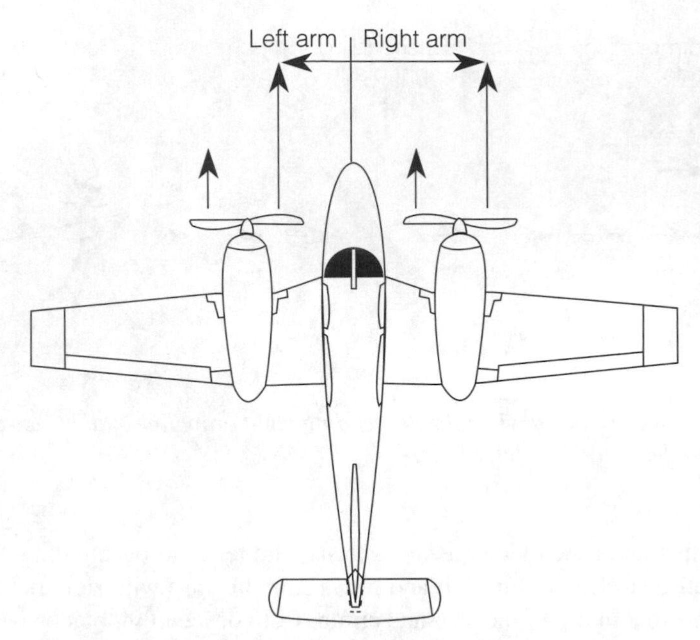

Fig. 1-12. *With a nose-high attitude, both descending propeller blades produce P-factor. This causes the center of thrust to shift to the right. When the right engine's thrust shifts, it gets farther from the airplane's center of gravity. The right lever arm increases while the left lever arm decreases.*

where a left engine failed, the operating right engine would produce a greater yaw force. This greater yaw force can only be overcome with a greater right rudder force. Greater rudder force can only come from greater airflow past the rudder, and greater airflow only happens with greater (faster) airspeed. So, the reason one engine is more critical than the other is because one engine's failure makes V_{MC} get faster.

In multiengine airplanes where both propellers turn to the right, the left engine is the critical engine. If the right engine quits, you still have a problem on your hands, of course. But the distance from the left engine's center of thrust to the airplane's center of gravity (the arm) is shorter and therefore the yaw force is weaker.

V_{MC} is improved as the P-factor of the left engine starts to shift the center of thrust toward the right. This is because as the thrust shifts to the right on the left engine, it narrows the gap to the airplane's center of gravity. The shorter the lever arm, the less yaw force produced, and the less force required to counteract.

Losing either engine is bad, but losing the critical engine is worse. Recovery from a situation where the critical engine has failed requires a faster airspeed because V_{MC} gets faster.

Picture a situation where a pilot is climbing out with an airspeed just slightly faster than the published V_{MC} (red radial line):

- The right engine fails and the airplane yaws to the right. The current airspeed allows for enough rudder effectiveness to offset this yaw and the pilot maintains control.
- The left engine fails and the airplane yaws to the left. The force of this yaw is greater because the yaw has greater leverage. The V_{MC} quickly gets faster than the published value. The pilot sees the airspeed indicator showing faster than V_{MC}, but the rudder force at this speed cannot overcome the magnified right-engine yaw force, and control is lost.

Attempts have been made to solve the critical engine problem. The best solution is to eliminate the critical engine altogether by making the propellers turn in opposite directions. Many general aviation twins have counterrotating propellers. As seen from the cockpit, the left engine turns clockwise and the right engine turns counterclockwise. This means that both descending blades are on the inside and therefore both are equally close to the airplane's center of gravity. If either counterrotating engine fails, the problem will be bad, but left and right are equally bad. V_{MC} remains the same regardless of which engine fails.

Maintenance on a counterrotating twin can be more expensive because certain parts are unique to a left-turning or a right-turning engine. Many parts cannot be interchanged, but the expense is worth it in a critical situation. You can't "buy" airspeed.

WEIGHT AND BALANCE

The loading of the multiengine airplane also has a large effect on V_{MC}. Pilots say weight and balance, and then refer to the calculation of center of gravity. Weight and balance must be separated and treated as singular topics for an understanding of V_{MC}.

Flying small airplanes, we are very mindful of the airplane's total weight. We are very careful not to flirt with that upper limit of "max gross." We learned that weight was bad and a lighter airplane was a safer airplane. Here again, V_{MC} is a paradox. A heavy airplane is actually better for V_{MC}. A heavy airplane is bad for performance, but remember that V_{MC} is a matter of control, not performance. The pilot must understand this important point: An airplane should not be flown heavier than its maximum allowable weight, but the closer that the weight is to the maximum allowable, the slower V_{MC} will become.

A heavy airplane is a stubborn airplane. The more mass the airplane has, the more inertia it will have. Think about this illustration: Two pilots are at the bowling alley playing on adjacent lanes. One pilot is bowling with a regular bowling ball (heavyweight) and the other pilot is bowling with a volleyball (lightweight). They both roll their respective balls at the bowling pins. While the balls are rolling down the lanes, a bowling alley employee opens a side door and a powerful crosswind sweeps across the lanes. The crosswind strikes the bowling ball, but the bowling ball is hardly affected. The same wind strikes the volleyball, and the volleyball's path is deflected into the gutter.

The wind was the same for both balls. We will assume that the balls were rolled with equal force and that balls were rolled down the center of each lane. Why did one drop in the gutter while the other did not? Weight. It takes a much greater force to move a heavy object.

When I see a pilot report about turbulence, I immediately look to see what type airplane the report came from. Severe turbulence reported by a Cessna 150 and severe turbulence reported by a Boeing 757 is a misrepresentation of the conditions because the two airplanes do not weigh the same. It takes a large force of air to throw an airliner around, but not very much force to toss a Cessna.

If one engine of a light twin fails, the entire airplane will start to yaw. A heavy airplane will resist any movement. It will take a greater yaw force to move the nose of a heavy airplane than a light airplane. What if a huge cruise ship lost the use of one propeller on one side of the ship? Would the ocean liner quickly begin a yaw-turn? No, the turn would be slow and labored. The ship is massive and therefore sluggish. What about a rowboat propelled with the same power as those cruise ship propellers, but only from one side? What will the rowboat do? It will turn immediately. The turn will take effect quickly because the rowboat is light and is pushed around easily.

Any yaw force caused by a failed engine will have less effect on a heavy airplane. When yaw has less effect, there is less force required to balance forces; therefore, less airspeed is required. V_{MC} is slower when heavier. When yaw is produced by a failed engine of a light airplane, however, things can happen faster. More rudder force and greater airspeed will be required. V_{MC} is faster when lighter.

Effect of CG

The location of the weight also affects the speed of V_{MC}. All pilots have calculated a weight and balance problem before. Take the weight of fuel, people, baggage, and the airplane, and determine where the focal point of all this weight rests. That point, the center of gravity, must then be placed in proper proximity to where the wing's lift is being generated in order to have proper aircraft stability. A center of gravity outside the safe range can be fatal, but for multiengine pilots, there is another reason for concern.

The "rudder arm" is shortened when the center of gravity moves to the rear of the airplane (Fig. 1-13). Remember that the pivot point of the airplane is the center of gravity. When the airplane yaws, it turns side to side around this crucial point. When the distance between the rudder and the center of gravity is long, the effective force will be magnified and therefore strong. (Archimedes would be proud.)

But as the gap narrows, the force gets weaker. If the rudder force is weak, it might not be able to overcome engine-out yaw forces at all. The airplane might be under the maximum allowable weight, and the center of gravity might be within the safe range, but V_{MC} gets faster with every increment of aft CG shift.

If an engine fails, the nose will begin to swing. The pilot will step on the rudder to stop the swing and straighten the nose, but with an aft CG, the application of rudder does not yield enough results. The engine yaw might overpower the weak rudder yaw, and aircraft control is quickly lost. The only way the rudder can avoid being overpowered is to have greater effectiveness through greater airspeed. V_{MC} goes up when CG moves back.

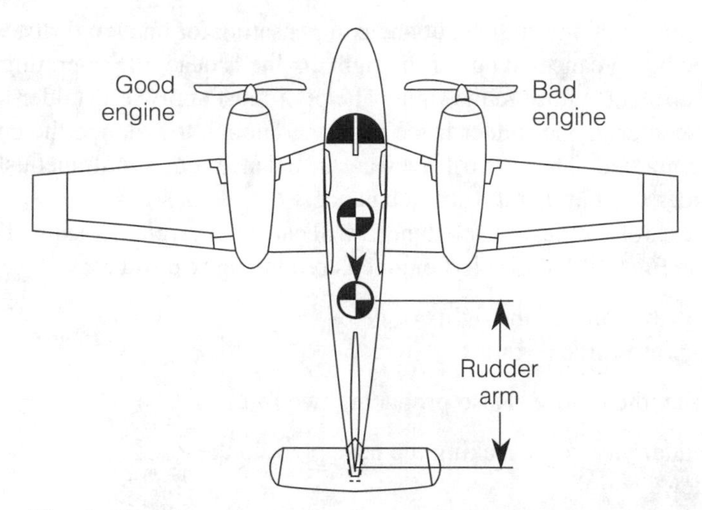

Fig. 1-13. *When the airplane's center of gravity moves aft, the distance (arm) between the pivot point and the rudder is reduced. This weakens the rudder's effectiveness.*

COORDINATION AND SIDESLIP

How the airplane is flown also has an impact on the speed of V_{MC}. Differences in airplanes and opinions have fueled a debate for many years on just how a multiengine airplane should be controlled when an engine has failed. At one time, it was preached that a pilot should never turn the airplane toward a dead engine. If the heading was north and a turn to east was required, the pilot would turn left—the long way to east—if the right engine had failed. Because FAR Part 23.149 specifies that the bank angle for V_{MC} testing be "not more than five degrees," many took the 5° bank as the absolute gospel.

These are the facts: Anytime the airplane's flight path is not parallel with the airplane's nose-to-tail line, drag will increase. Anytime a slip occurs, the rudder effectiveness will suffer. Both factors, drag and rudder effectiveness, have a big impact on the speed of V_{MC}. Anything that tampers with drag or the rudder also changes V_{MC}; therefore, the pilot must understand what is taking place.

If the pilot flies with the wings level and the ball of the inclinometer is in the center while one engine is failed, V_{MC} will get faster. All previous single-engine training has taught that "ball in the center" is a good thing. Your flight instructor spent many hours trying to convince you that staying coordinated during turns and takeoffs, especially during stalls, was vital. But when one engine fails, once again V_{MC} is a paradox. To get best results with V_{MC}, you must go against previous single-engine training. The slowest speed that can be obtained for V_{MC} takes place at *zero sideslip*.

To understand what zero sideslip means, the pilot must first understand what a sideslip is and why it is being produced. A slip takes place when the airplane does not travel in the direction that the nose is pointed. This can happen when flying through a crosswind, but for this discussion wind will not be a factor.

When flying with one engine out and compensating for uneven thrust with rudder, two forces are being canceled out. If the right engine is dead, the operating left engine will yaw the airplane's nose to the right. The pilot must apply left rudder to swing the nose back to center. If the rudder force is strong enough to balance the engine force, control is maintained. These two forces are in balance, but simultaneously there are two other forces present that are not in balance.

The "other" forces that are unbalanced will cause the airplane to slip. The right engine is dead in Fig. 1-14A. The left engine is producing two forces:

- Engine yaw force to the right
- The forward force of thrust

Meanwhile, the rudder is also producing two forces:

- The rudder yaw force making the nose pivot to the left
- A "sideways lift"

The two yaw forces (labeled 1 and 3 in Fig. 1-14A) offset each other and cancel out, but the left propeller's forward thrust force and the rudder's sideways lift force do not offset and therefore do not cancel out.

What is *sideways lift*? The vertical stabilizer and rudder combination acts just like the wing and aileron combination. When an aileron is deflected down, the airflow on the wing's upper camber has a greater distance to travel. This additional distance causes the air molecules to speed up. When the airflow gets faster, the drop in pressure is greater. This drop in pressure creates more lift and the wing rises.

The same thing happens with the rudder. When the rudder is deflected, airflow velocity increases in the area that the travel distance lengthens. The faster that the molecules of air travel, the lower pressure that is produced. When the left rudder pedal is pushed, the rudder's trailing edge moves left. This produces a greater camber on the right side of the vertical stabilizer/rudder combination. A lower pressure then forms on the right side of the stabilizer. The entire tail section then is drawn into the area of low pressure. When the tail moves to the right, the airplane pivots on the center of gravity and the airplane's nose moves left. The sideways lift is a differential of pressure on opposite sides of the vertical stabilizer.

Thrust pulls the airplane forward; rudder force pulls the airplane sideways. The two forces acting on the same body come to a compromise: a slip. The slip is actually the resultant vector of the two forces (Fig. 1-14B). Thrust, as you might have already determined, is much stronger than sideways lift; therefore, the resultant is more forward than to the side. But as sure as the airplane is moving forward, it is also moving sideways, and anytime that the airplane is not completely traveling forward, there will be greater drag. The greater drag is a result of the relative wind striking the airplane from the side rather than from straight on. So, sideways lift causes drag and drag causes V_{MC} to get faster.

Also, the slip damages the rudder's ability to do its job. If the travel path of the airplane is not straight, then the airplane's relative wind is not straight. Figure 1-15 shows the airplane in a sideslip to the right. The right engine is dead. The resultant vector of

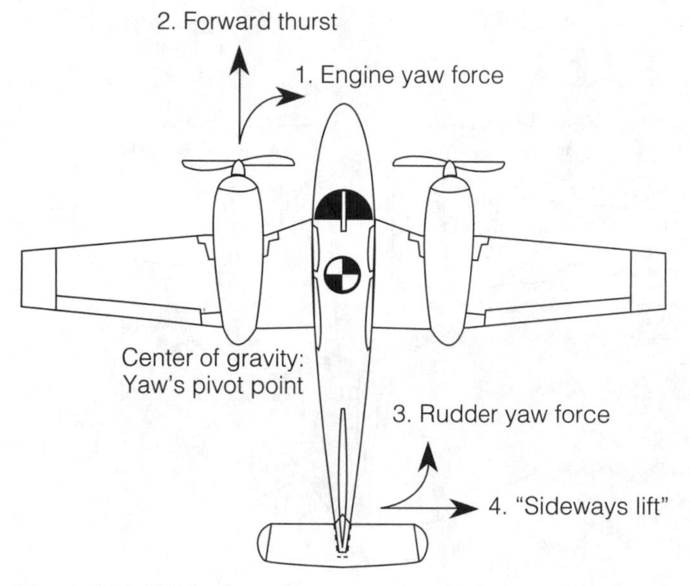

Fig. 1-14A. *With the right engine inoperative and the pilot holding left rudder, four forces go to work on the airplane.*

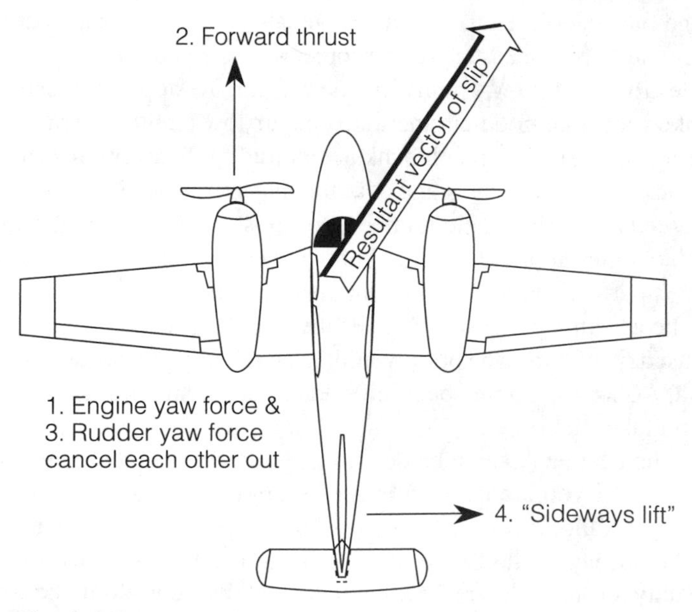

Fig. 1-14B. *A side-slip is the result of forward thrust and the rudder's "sideways" lift joining forces.*

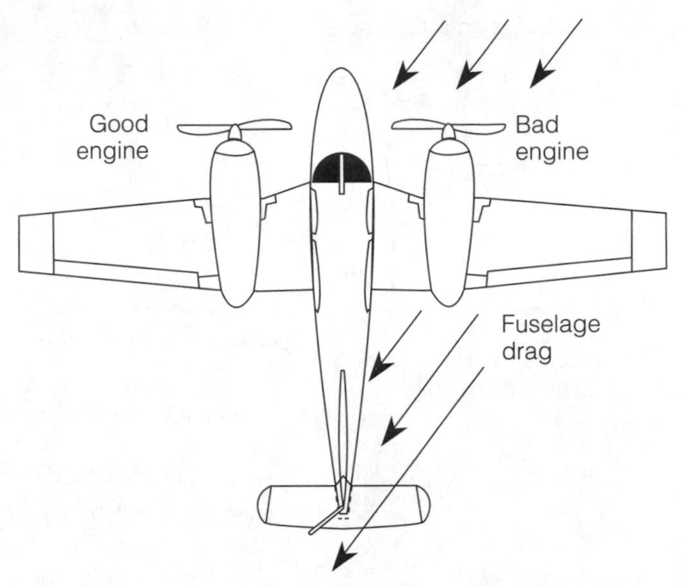

Fig. 1-15. *When the airplane is in a slip, the relative wind produces more drag by striking the fuselage on the side, and rudder effectiveness is lost because the relative wind crosses the rudder with a more parallel angle.*

thrust and sideways lift are producing a travel path that is forward, but also to the right.

Look at the way the wind strikes the deflected rudder. The greatest force is produced when the wind hits an object head-on. In this situation, the wind is almost parallel to the plane of the rudder. This makes the rudder force weaker and drives V_{MC} faster.

So, a sideslip is bad for V_{MC} and airplane control. To stop the sideslip, the airplane must be banked into the good engine until the airflow again is parallel with the airplane's nose-to-tail line. How much bank is required? Whatever bank that is necessary to stop the sideslip, which is the same as achieving zero sideslip.

Five degrees is not the magic number. For most twins, the number is closer to 3° bank. V_{MC} is lowered approximately 3 knots for every degree of bank closer to zero sideslip. If the "perfect" angle were 3°, but the pilot elected to fly with wings level, the result would be a 9-knot increase in V_{MC}. If the pilot forgot about zero sideslip and automatically used 5° of bank, the pilot would be penalizing performance by raising V_{MC} an unnecessary 6 knots. These speed differences seem small, but even 1 knot under V_{MC} is fatal on takeoff.

How can the best bank angle be determined? I recommend an old-fashioned way: Use a yaw string. If you are training in a multiengine airplane without a yaw string, your training is lacking. It is easy to do. Take some heavy string and attach one end to the nose of the airplane. Make sure that the string is attached exactly on the airplane centerline. Many airplanes have a seam in the metal that runs along the centerline back to the windshield. Use this seam as the reference line. If there is no convenient seam, mark the centerline with a piece of tape.

While in normal, two-engine flight, the string will flutter in the relative wind, but it will be straight along the centerline (Fig. 1-16). Now, reduce the power on the left engine to simulate zero thrust and initially fly the airplane with wings level. You will see that with the wings level, the yaw string does not lay along the centerline, but the ball is in the center (Fig. 1-17).

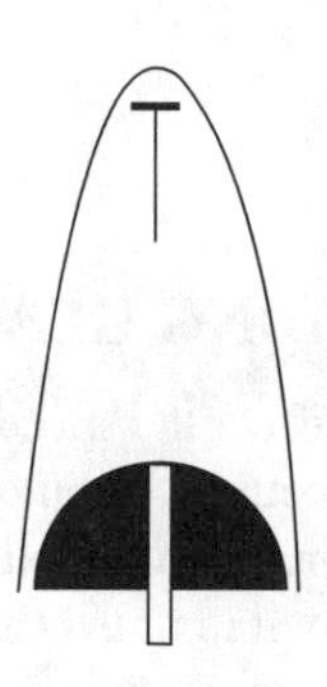

Fig. 1-16. *As seen from the left pilot's seat, the yaw string is standing straight back with both engines operating. The fuselage centerline seam is your reference with the string. The tip of the string will flutter due to the faster, two-engine speed.*

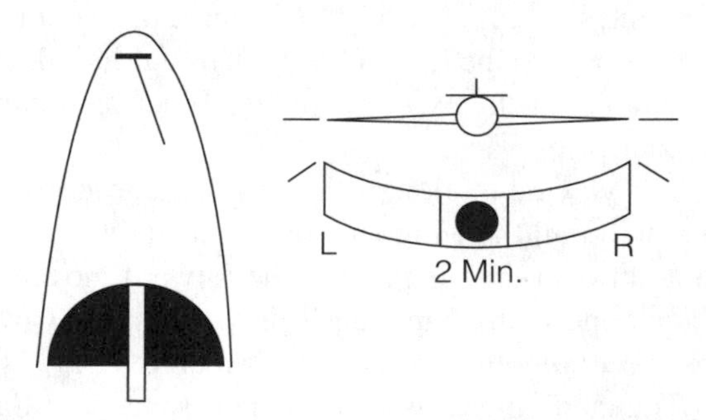

Fig. 1-17. *The left engine is inoperative, and the wings are level. Side slip is noted by the position of the yaw string, right of the centerline. The ball is held in the center of the inclinometer.*

Remember that anytime the yaw string is not on the centerline, the relative wind is coming from the side and drag is increased on the fuselage. Anytime the yaw string does not stretch across the centerline, the rudder is less effective. Both factors tell you that anytime the yaw string is not on the centerline, V_{MC} is faster.

Now, bank into the good engine until the yaw string is on the centerline (Fig. 1-18). When the string centers, the airplane's bank angle has achieved zero sideslip. Now over-bank and watch the string move past the centerline; V_{MC} starts to increase again.

A curious thing happens to the ball during the yaw-string exercise. While flying straight and level, the yaw string is not on the centerline, but the ball is centered (Fig. 1-17). When the yaw string is on the centerline, the ball is about halfway out of center in the direction of the good engine (Fig. 1-18). The paradox of V_{MC} is proven again: Straight and level with the ball centered is not as good as a bank with the ball halfway out of center.

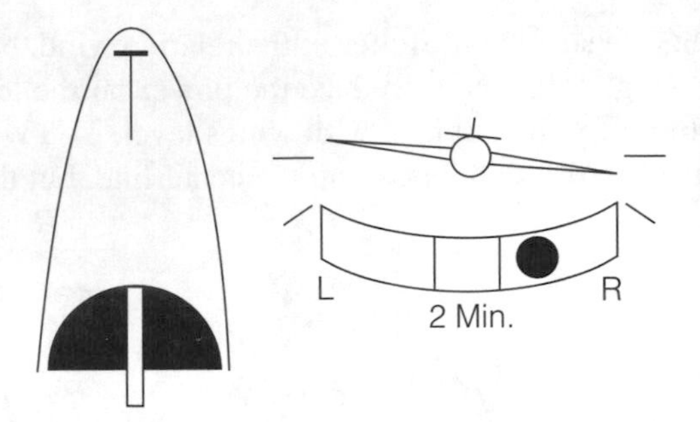

Fig. 1-18. *The airplane has been banked toward the right, the direction of the good engine. The airplane is now flying with zero side-slip, which is noted by the string on the centerline and the ball out of the center, also in the direction of the operating engine.*

FLAPS AND LANDING GEAR AND V_{MC}

Flaps and landing gear affect the speed of V_{MC}. Up until now the factors affecting V_{MC} have involved an asymmetrical situation where something happened on one side of the airplane that did not happen on the other side: engine out. Flaps and landing gear are supposed to go up or down in unison and no asymmetrical situation occurs. Lowering flaps and landing gear in many twin-engine airplanes can reduce the speed of V_{MC}.

Extended flaps or landing gear act as stabilizers. Jet fighter planes often require drag chutes to stop. Usually a parachute pops out the back and the air resistance of the chute helps stop the fighter. Wing flaps produce drag, so imagine that your airplane has two small drag chutes, one per wing. When simultaneously deployed, the chutes would pull each wing back.

The yawing motion, which is so dangerous to multiengine pilots when one engine quits, will pull the wing with the good engine forward; however, if the flaps are down and additional drag is present, the forward movement of the wing as it pivots on CG will be opposed by flap drag. This dampens the yaw. Less yaw means less rudder force is required; therefore, less speed is required: V_{MC} is lowered.

The landing gear works much the same way. Sailboats have a long keel that sticks down deep underneath the hull of the boat. The keel's job is to grip the water and provide stability to the boat. When the landing gear and tires stick down into the relative wind, they grip the air and will resist yaw.

So, lowering wing flaps and landing gear can slightly improve airplane control, but unfortunately it will devastate performance. The extra drag of flaps and gear will bring the airplane down. Trying to pull flaps and gear through the air with only one engine is a losing battle. For this reason, a pilot should not rely upon the application of flaps and landing gear to solve V_{MC} problems, except under extreme conditions.

Obviously, anything that causes more drag on one side of the airplane while flying on one engine is terribly bad. It would be a disaster to have only one wing flap lowered. If a pilot making an approach to landing got too slow and too close to V_{MC}, then hit the flap control and only a single flap went down, the result would adversely affect V_{MC}. As the split-flap condition developed, especially if the flap went down on the dead-engine side, a great yaw force would develop and V_{MC} would skyrocket past the approach speed.

Before takeoff, run the flaps through their paces. Make sure that they go up and down together. Always keep your hand on the flap control while the flaps are in transition, or at least until you are sure they are lowering evenly.

One other caution: When operating with only one engine and flaps down, be ready for the good engine's wing to rise if you go full power. Airflow from off the propellers streaks back across the wing flaps of conventional twins. This *accelerated slipstream* will produce faster airflow over the camber.

When one engine has quit, the accelerated slipstream will only exist on one side and the spanwise lift pattern will be asymmetrical. In other words, the wing with the operating engine will have more lift and this starts a yaw-turn that will raise V_{MC}.

This particular situation might occur when a single-engine go-around is attempted. A single-engine go-around is a very bad idea. The ability of the operating engine to pull you back to a safe altitude, together with the faster V_{MC} during a flap-down application of full power, makes this very dangerous. It might be better to land in the grass beside the runway.

SITUATIONAL AWARENESS

Now that the theory has been explained, it is time for some practical application. You are flying a conventional twin-engine lightplane. Your current airspeed is just faster than the published V_{MC} and one engine has already failed. How will the following situations affect the speed of V_{MC}? Do any of these situations make the pilot "less safe?"

1. Density altitude is 8,000 feet.
2. The dead engine's constant-speed propeller control cable breaks.
3. The airplane is flying with full fuel, all seats filled with passengers, and full baggage compartments. It is at maximum gross weight.
4. A full set of golf clubs is in the aft baggage compartment.
5. A circuit breaker pops leaving both wing flaps stuck down.

The question "Are you safer or less safe?" is too simple for these situations because what is good for V_{MC} is often bad for performance. What is "safer" for V_{MC} might at the same time be "less safe" from a performance standpoint. Because the question calls for a speed just faster than V_{MC}, we will consider "safer" to mean a margin of speed that is faster than V_{MC}.

What about situation number 1? The speed of V_{MC} will definitely be slower. The power output of the operating engine will be severely reduced while the engine gasps for air. Many pilots just do not completely understand the concept of density altitude. Some pilots can even calculate density altitude on a calculator-type flight computer, but they have no idea how to apply the information.

Can you be at 2,000 feet MSL and have an 8,000-foot density altitude? Of course. Can you be at 2,000 feet and have a "negative" density altitude? Of course. Multiengine flying requires pilots to get back to basics. The importance of the basics is amplified when applied to multiengine flying. When the engine produces less power and less yaw, there is less need for airspeed to cross the rudder; however, an 8,000-foot density altitude is "less safe" for performance, especially if you are attempting to climb over a mountain ridge.

The answer for situation number 2 depends upon the airplane's individual systems, which are covered later. But for now, assume that when the cable that operates the propeller's blade angle breaks, you lose the ability to pull the propeller into the feathered position. Being unable to prevent the propeller from windmilling will increase drag and increase the speed of V_{MC}.

In situation number 3, the airplane will have poor performance characteristics such as rate of climb, but V_{MC} will be slower. Never try to justify flying at or heavier than maximum gross weight by using the fact that V_{MC} is improved.

Number 4 is interesting. A full set of golf clubs is very heavy, which can be dangerous if you magnify this weight by placing the clubs in the aft compartment. Even if the CG shift caused by the aft weight is still in the safe CG range, V_{MC} is going to be faster. The distance between CG (pivot point) and center-of-rudder force is reduced and weakened. When one engine quits and you step on the rudder to overcome the yaw, you will run out of rudder before the job is done. Consider putting the clubs in the back seat or renting clubs when you get there.

Number 5 is safer because some small degree of extra stability will slow V_{MC} slightly. Remember that if this happens down low, maintain control and land on something flat. Do not ask the airplane to do something that is impossible and lose airspeed below V_{MC} in the attempt.

You can certainly make up other V_{MC}/performance scenarios. The key is to completely understand what will affect what. This closing table summarizes everything covered in the chapter. Do not simply memorize the table's information; only through understanding the information will you be able to safely fly a multiengine airplane.

What changes the airplane's minimum control speed?

Factor	**Higher V_{MC} (bad)**	**Lower V_{MC} (good)**
Prop rotation	Critical engine; Props turn same way	No critical engine; Counter-rotating props
Prop condition	Windmilling	Feathered
Power produced	High; Low altitude; Cold temperature	Low; High altitude; Hot temperature
Total weight	Light	Heavy
Center of gravity	CG aft	CG forward
Coordination	Ball centered; Side slip	Ball ½ out; Zero side slip
Bank	Wings level	Banked to good engine

2 Multiengine takeoff and landing

TAKEOFF IN A MULTIENGINE AIRPLANE IS MORE HAZARDOUS THAN ANY other maneuver. When the airplane has reached a safe altitude and speed, having a second engine is great, but in the transition zone between slow and fast, two engines double the dangers. Multiengine pilots must plan every takeoff with the expectation that one engine will fail. If an engine fails during takeoff, it will be a surprise to the pilot; how the pilot reacts should not be a surprise.

SPEED MILESTONES

As a normal takeoff progresses, the plane and pilot will pass definable points of decision that I call "speed milestones." The proper action taken by the pilot when one engine quits on takeoff changes according to what milestones have been passed. In order to do the right thing, the pilot must understand that different speeds produce different reactions from the airplane (Fig. 2-1).

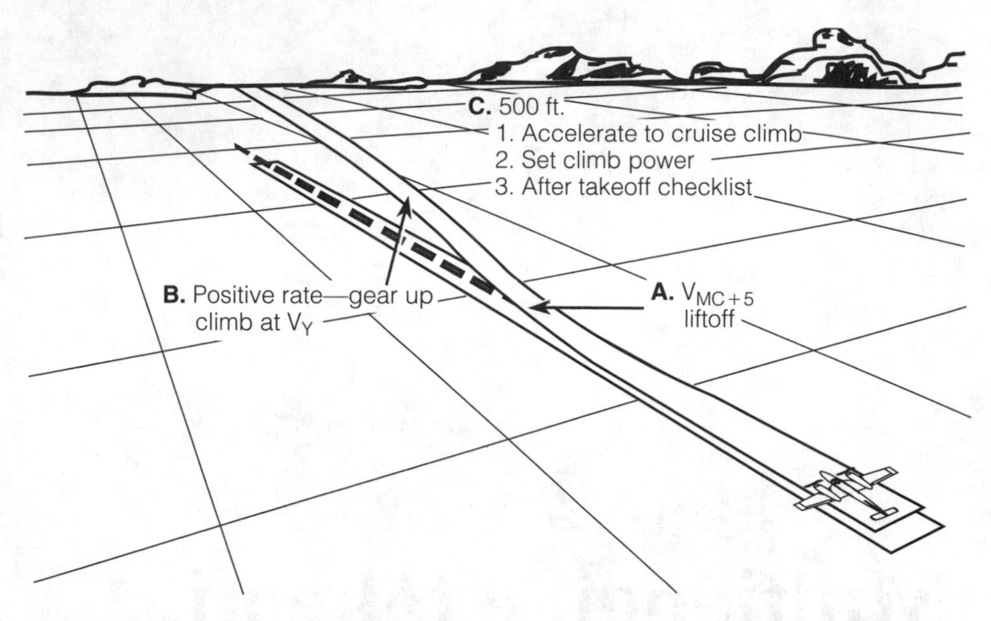

Fig. 2-1. *Normal multiengine takeoff.*

Zero

The first speed milestone is zero knots. Before letting the airplane move down the runway, the pilot must be absolutely certain that the airplane is ready. It is a good practice to align the airplane with the runway centerline and then pause for a few seconds. While holding the brakes, run the engines up to takeoff power, and make sure that the engine instruments show green. Verify that the manifold pressure and RPMs are acceptable. Be sensitive to any shudders or shakes that are unusual.

Normal pretakeoff runups are done with less than takeoff power, so this is the first time you can check the airplane at this higher power setting. If there is any indication of a problem, you would rather discover it while standing still; you can turn your full attention to engine observation. When you let go of the brakes, you will be too busy with steering, crosswinds, and other distractions to carefully watch the engine instruments.

Taking the runway, but delaying takeoff for an engine test does have its hazards. The controllers in the tower do not like pilots to make unannounced delays on an active runway. Observe the traffic load by listening to the tower or unicom frequency before you call "Ready for takeoff." This will build your airport situational awareness. If the traffic is heavy, with one airplane after another landing, it will be best to advise the tower that you will need some time on the runway before you begin the roll. Just say, "November . . . is ready for takeoff, and we will need a 15-second delay on the runway."

The controller should know exactly what you are doing and might be able to space traffic accordingly. At uncontrolled airports, an unexpected delay could cause another pilot to go around. Protect yourself by checking that everything is working properly before letting the airplane move, but be courteous as well.

V_{MC}

When V_{MC} is reached and surpassed, another speed milestone passes. The minimum control speed must be observed and the airplane held on the ground through this speed. It is a good idea to observe the airspeed indicator for these milestones to pass. At the very beginning of the takeoff roll, the airspeed indicator will not move because most indicators do not go as low as zero. The plane must be into the takeoff roll several seconds until the airspeed indicator starts to move. This is when I say out loud "airspeed's alive" even if I am alone. This verifies that the airspeed indicator really works, which is a nonspecific speed milestone.

V_R

The next speed milestone is reached at the liftoff speed or V_R. V_R stands for rotation and is the speed when the pilot pulls back on the wheel and rotates the nose up and off the ground. The V_R speed can never be safely slower than V_{MC}. If V_R were slower than V_{MC}, and at this speed an engine failed, the pilot would be unable to control the airplane. A safe recovery would be unlikely at such a low altitude. Allowing the airplane to become airborne at a speed slower than V_{MC} should never be acceptable to the pilot.

Check the recommended V_R speed in the operating handbook. Ordinarily, V_R is equal to V_{MC} plus 5 knots ($V_R = V_{MC} + 5$). The zero speed milestone and the V_R milestone form the first takeoff "decision zone." Ask yourself: "If an engine fails now, do I stop or continue to take off?" Between zero and V_R, the decision is easy: Stop.

If anything happens in this speed range that makes you suspicious about the condition of the airplane, immediately reduce power on *both engines*. Pilot examiners predictably present engine problems to multiengine checkride applicants during this decision zone. They not only want to see that you will abort the takeoff while between zero and V_R, but that you will bring the power back on both engines.

The examiner might say, "You have low oil pressure in the left engine." If the multiengine applicant pulls back the throttle on only the left engine, the checkride will be over. If only one engine is reduced, the other engine at full power will likely pull the airplane off the centerline and into the runway lights. You do not have time during this part of takeoff roll to do much troubleshooting. Just pull everything back, steer with your feet, bring the airplane to a stop, and inform the control tower or make an announcement on the CTAF.

V_{YSE}

Decisions become more complicated after liftoff. V_{YSE} is the next milestone speed—the best rate of climb speed using only one engine. V_{YSE} is marked on the airspeed indicator with a blue radial line. It should probably be called the best one-engine performance speed because sometimes the best performance is still a descent. Nevertheless, V_{YSE} is a milestone of significant importance.

If an engine fails after reaching V_{YSE}, the pilot is already at the best speed for that predicament. In this case, the pilot could move attention to identifying the failed engine and reducing drag, rather than changing speeds. V_{YSE} would equal V_R in a perfect world.

If it were possible to safely hold the airplane on the ground until it accelerated to V_{YSE}, then an engine failure would occur at best performance speed. Holding the airplane on the ground is unsafe because wheelbarrowing might occur. If a pilot holds the wheel forward too long in an attempt to increase liftoff speed, the main wheels might lift off when the wings have enough lift and the nose wheel remains on the ground. The airplane would look just like a wheelbarrow being pushed along.

The airplane might start to "buck" as the wings pull the airplane up and the pilot fights to keep it down. It is not a new problem. So many takeoff accidents have been attributed to wheelbarrowing that the FAA published an advisory circular in 1968 (AC 90-34).

The greatest danger is a combination of wheelbarrowing and crosswinds. If the airplane were allowed to ride up on only the nosewheel, a crosswind could easy pivot the airplane into the wind like a weather vane and directional control would be lost.

The airplane should be allowed to become airborne at $V_{MC} + 5 = V_R$ and then accelerate when free of the ground to V_{YSE}. While the airplane is accelerating through the zone of decision between V_R and V_{YSE}, the proper pilot action to take in the event of an engine failure depends on several factors: density altitude, runway length, wind, and obstructions to climb over.

If an engine quits in this zone (between V_R and V_{YSE}), the best option is to pull both throttles back, and land straight ahead on the remaining runway. Never retract the landing gear while there is runway ahead and the airplane is accelerating through this zone. The landing gear does produce drag, but if the best decision is to land anyway, drag is not a big problem. Retract the landing gear only after reaching V_{YSE} or reaching a position where no landing spot is available ahead.

If strong yaw occurs in this zone and plenty of runway is available, retard both throttles. It can happen so fast that there will be no time to think about which engine has failed. Even if you do figure out which one has failed, reducing power on the failed engine does not help. The good engine would still be producing takeoff power, and this would continue the yaw force. If an engine quits, pull both throttles back and correct for yaw and land.

A multiengine pilot must understand the airplane's abilities and limitations in this speed range of critical decisions; otherwise the pilot might inadvertently ask the airplane to do something that it cannot do.

THE CLASSIC MISUNDERSTANDING: MULTI VS. SINGLE

Who is safer when one engine quits just after takeoff? The single-engine pilot or the multiengine pilot? Picture a purely hypothetical situation where two pilots face a takeoff on parallel runways. Takeoff conditions are identical for both airplanes. One airplane has two engines and the other has one engine.

Both pilots add takeoff power and the airplanes begin to roll. Both accelerate at the same speed. Both rotate and break ground at the same place. Each pilot has the same amount of runway ahead with trees at the end. At this moment, each airplane loses an engine.

The single-engine pilot's decisions are already made: Control the airplane until it hits the ground. Without an extra engine to complicate things, the single-engine pilot turns his attention to the only situation left, a forced landing. It never enters the single-engine pilot's mind to attempt to stay in the air.

The multiengine pilot might entertain the idea that a forced landing might be avoided because the airplane has an extra engine. Staying in the air and avoiding a hazardous forced landing is the best option, but just because a pilot thinks that this is possible does not mean that the airplane is capable of staying in the air.

Certain speed and altitude situations will prevent the multiengine airplane from climbing to a safe altitude or even maintaining altitude. Just like the single-engine pilot, the decision might already be made—a forced landing. If the forced landing option is not realized, the multiengine pilot might waste valuable time and airspeed trying to milk the airplane higher; hence, the airspeed falls below V_R speed as the pilot aims the nose to the sky, then the airplane falls below V_{MC} and control is lost.

There are times when the multiengine pilot must accept the fact that a climb is not possible and a forced landing is inevitable, even with one engine operating at takeoff power. The pilot has no good choices, but the best option left is to land ahead, under control.

The other option is a trap. The operating engine might deceive the pilot with the promise of extra safety, but airspeed is lost while chasing the climb and the airplane lands (crashes) out of control. The old saying is true in this speed range: "Below V_{YSE}, if one engine quits, the operating engine simply takes you to the scene of the accident."

Dealing with differences

Remember that every airplane and every takeoff is different. Certain airplanes will pull out of a situation when other airplanes will crash. Density altitude conditions will allow an airplane to climb to safety one day, but other days the exact same airplane cannot climb to safety. I could not write down an exact engine failure procedure that would guarantee safety between the milestones of V_R and V_{YSE}.

The only way to ensure a safe getaway on one engine is to lower the airplane's nose until an adequate speed is obtained and then start a climb at V_{YSE}. The problem with that statement is that it calls for a trade-off of energy: airspeed gained due to altitude lost. It is assumed that the pilot has no altitude to give if the airplane has just left the ground; therefore, the pilot is in a catch-22. The pilot needs airspeed but cannot get any without altitude, and he cannot get altitude without some additional airspeed.

Milestone planning

The best way to avoid all these problems is to plan the takeoff milestones. Decide prior to takeoff just what you will do if an engine fails between V_R and V_{YSE}. Take all factors into consideration, make a plan, and then stick with the plan. This way, if an engine fails, you will not waste time, airspeed, and altitude during a state of indecision. The decisions up until this speed range have been easy.

Speed range	*Pilot action when one engine quits*
Zero to V_{MC}	Retard both throttles and brake (easy decision).
V_{MC} to V_R	Retard both throttles and brake (easy decision).
V_R to V_{YSE}	Optional (tough decision). Option 1: Retard both throttles and land ahead. Option 2: Lower nose to gain V_{YSE}, then climb to safe altitude.

Option 2 might not be possible. If option 2 is not a true option, it is better for the pilot to accept option 1 early. By making an honest evaluation of the airplane's single-engine climb performance, the pilot should commit to either Option 1 or Option 2 before ever taking the runway for takeoff (Fig. 2-2). The pilot should say, "If I lose an engine before reaching 88 knots (V_{YSE}), then I put it back down in the smoothest place I can find."

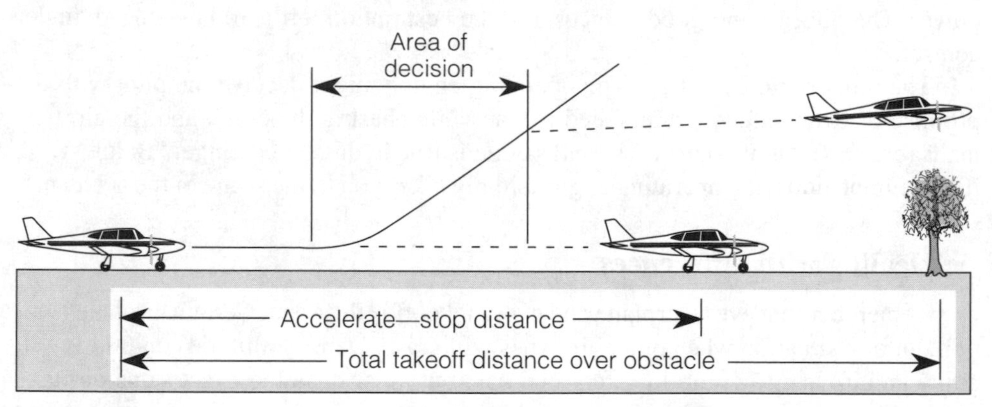

Fig. 2-2. *The area of decision between V_R and V_{YSE}.*

The 88-knot V_{YSE} is only an example, but it reflects what should be going through the pilot's mind when anticipating takeoff. If an engine quits at 79 knots while airborne, the pilot has already made the tough decision to land ahead. No time will be wasted trying to figure out what is better. Multiengine pilots must know the airplane's V speeds and milestones by heart in order to make safe decisions.

Making safe decisions

The pilot should consult several aircraft performance charts during preflight planning. The performance charts provided by the aircraft manufacturer usually are best-case numbers because they were derived by an experienced test pilot flying a new airplane. Pilots should know how to use the performance charts to get the numbers, but understand that actual performance will most always be worse than the chart numbers predict.

Start with the ground-roll graph. These numbers predict the distance, in feet, from initiation of the takeoff roll until liftoff. Figure 2-3 is a typical ground-roll graph. These graphs require a steady hand, and I usually close one eye so the lines do not run together. Use a plastic overlay, and draw lines on the plastic rather than the chart. If you draw the lines on the chart and then erase the next takeoff calculation, the graph will not last very long.

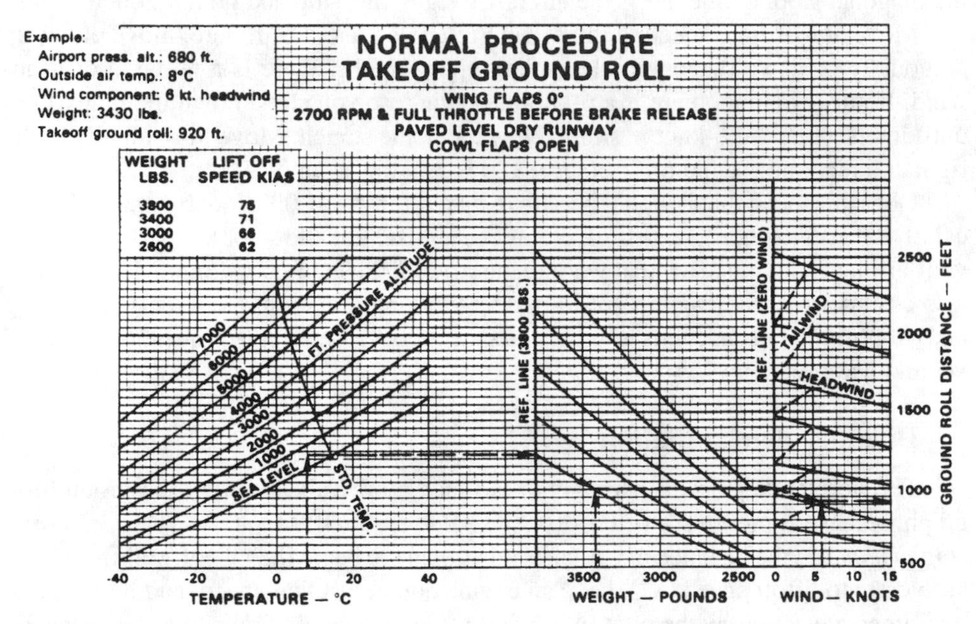

Fig. 2-3. *Normal takeoff ground-roll performance chart. (Copyright Piper Aircraft Corporation. Used with permission.* ***WARNING:*** *This performance chart is presented for illustration purposes only and is not intended to predict the performance of any specific airplane. Piper Aircraft Corporation is not involved with publishing this book and does not endorse any statements made herein.)*

To get any information from this maze, you first must determine the following variables:

- Pressure altitude of the airport.
- Temperature in degrees Celsius.
- Weight of the airplane.
- Wind on the runway.

The manufacturer has provided a sample on this graph: pressure altitude is 680 feet, temperature is 8°C, aircraft weight is 3,430 pounds, and a headwind component is 6 knots. Starting at the lower left on the temperature scale, trace up the graph at 8°C until you hit the 680-foot pressure altitude line. Surprise! There is no 680-foot altitude line. This is where the eye squinting comes in. You must estimate where 680 feet should be between the sea-level and 1,000-foot lines.

This rough estimating of position always scares me because a mistake here will change the answer to the problem, and it is upon this answer that safe takeoff decisions will be made. From the estimate of 680, make a 90° right turn and move horizontally to the first reference line. From the reference line, move diagonally with the sloping weight lines. From the base of the graph, find the airplane's weight and start up. When the diagonal sloping line meets the aircraft-weight line, stop and go horizontal again.

Move across until meeting the second reference line. From here, move either up diagonally or down diagonally depending upon whether there is a tailwind or headwind. In this example, there is a 6-knot headwind, so you slope down the index lines until intersecting the 6-knot position. Now the home stretch. Move horizontally to the right and read the ground roll distance, 920 feet in this example.

Does this mean that it would be safe to take off from a 1,000-foot runway? No. Recall that these numbers are best-case numbers and do not allow for the performance of your airplane and mistakes that were made reading the confusing graph. The 920 feet also assumes that the engines are in perfect working order and are evenly producing takeoff power. What happens when one engine quits? To answer that question, we need yet another graph.

Distance determination

The accelerate-stop distance graph is much more valuable than the ground-roll graph because it presents a better distance value required for safety. The *accelerate-stop distance* is the distance that it takes an airplane to begin the takeoff roll from zero, accelerate to liftoff speed (V_R), have an engine quit at just that speed, and be followed by immediate action by the pilot to retard throttles and safely brake to a stop—ideally on the runway. Broadly defined, the distance required to go from zero to V_R and then back to zero is usually more than twice as long as the calculated ground roll.

Figure 2-4 is a typical accelerate-stop distance graph. The manufacturer provides a sample problem with virtually the same conditions as before: 680 feet pressure altitude, 8°C, 3,430-pound airplane, and a 5-knot headwind component. Perform the graph calculation as before. The accelerate-stop distance answer is 2,050 feet of runway; the ground roll from the previous graph was 920 feet.

Multiengine pilots must look at these numbers realistically. Provided that the graph numbers are correct, a pilot could get into the air with only 1,000 feet of runway, but if anything goes wrong, the pilot will need 2,100 feet to get stopped again. The 2,100-foot number also assumes that the pilot recognized the problem of engine failure at the instant it occurred, and in the next instant the pilot pulled throttles and applied brakes simultaneously.

If the pilot is the least bit surprised, or panic prevents immediate action, that 2,100 feet will grow to 2,500 feet quickly. What if the airplane becomes airborne,

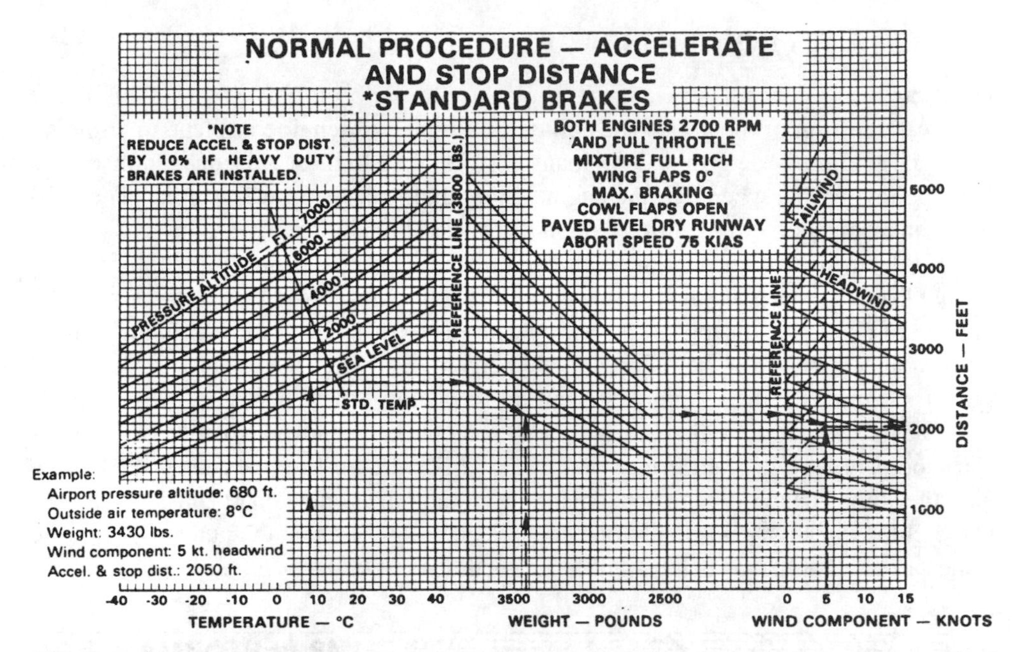

Fig. 2-4. *Accelerate-stop distance chart. (Copyright Piper Aircraft Corporation. Used with permission.* ***WARNING:*** *This performance chart is presented for illustration purposes only and is not intended to predict the performance of any specific airplane. Piper Aircraft Corporation is not involved with publishing this book and does not endorse any statements made herein.)*

then an engine quits while the airplane speed is between V_R and V_{YSE}? If the decision is to pull throttles and land, then brake, the distance needed to stop can grow to 4,000 feet or farther. The decision becomes tougher when the runway is only 3,500 feet long.

Imagine yourself in the pilot's seat, no more than 35 feet in the air after takeoff, and one engine has just failed. Airspeed is between liftoff and best single-engine climb speed. It would take 4,000 feet of runway to get stopped, but the total runway length is 3,500 feet. Only 1,500 feet of runway, then a fence, lie ahead of you. What should you do?

Option 1: Attempt to climb to safety.

Option 2: Land and brake.

If you knew for a fact that a climb to a safe altitude was possible, then you should add full power to both throttles (this ensures that the good engine is at full power) and then clean up the airplane by bringing up the landing gear, retracting flaps, and feathering the prop on the failed engine.

If you knew for a fact that a climb was not possible, then you should bring both throttles back, land on the remaining runway surface, and brake hard into the fence. It is always better to hit a fence while decelerating on the ground than to hit trees or terrain while accelerating (downward) in the air.

TOUGHEST QUESTION

How can you know for a fact if the airplane will climb on one engine? This is the toughest question of the book. The ability for a light twin-engine airplane to climb on only one engine is determined by many factors. The biggest misconception comes from believing that when half the engines stop, then half the power is gone. The ability of an airplane to climb is based upon its excess-thrust horsepower.

Excess refers to the airplane's power beyond what is required to maintain level flight. The airplane must have a surplus of thrust to climb. When one engine fails, 50 percent of the power disappears, so it would be easy to assume that the airplane's climb performance would also be reduced by 50 percent. But when asymmetrical thrust and drag are figured in, the actual performance loss is closer to 80 percent.

Can any airplane climb with a sudden 80-percent loss of thrust? Not many light twins can. Compare the two performance charts (Figs. 2-5 and 2-6). The first chart considers the airplane climb capability with both engines operating; the second chart considers the airplane climb capability with only one engine operating. The manufacturer provides the same sample problem as before. Under the conditions of 680 feet pressure altitude, 8°C,

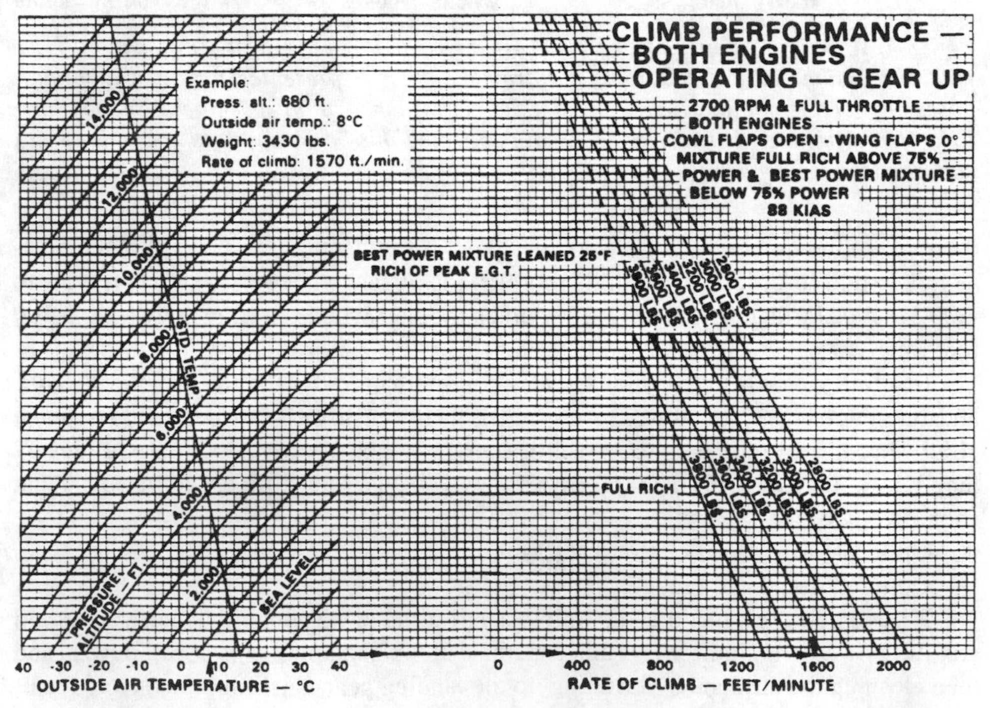

Fig. 2-5. *Climb performance chart: both engines operating. (Copyright Piper Aircraft Corporation. Used with permission.* ***WARNING:*** *This performance chart is presented for illustration purposes only and is not intended to predict the performance of any specific airplane. Piper Aircraft Corporation is not involved with publishing this book and does not endorse any statements made herein.)*

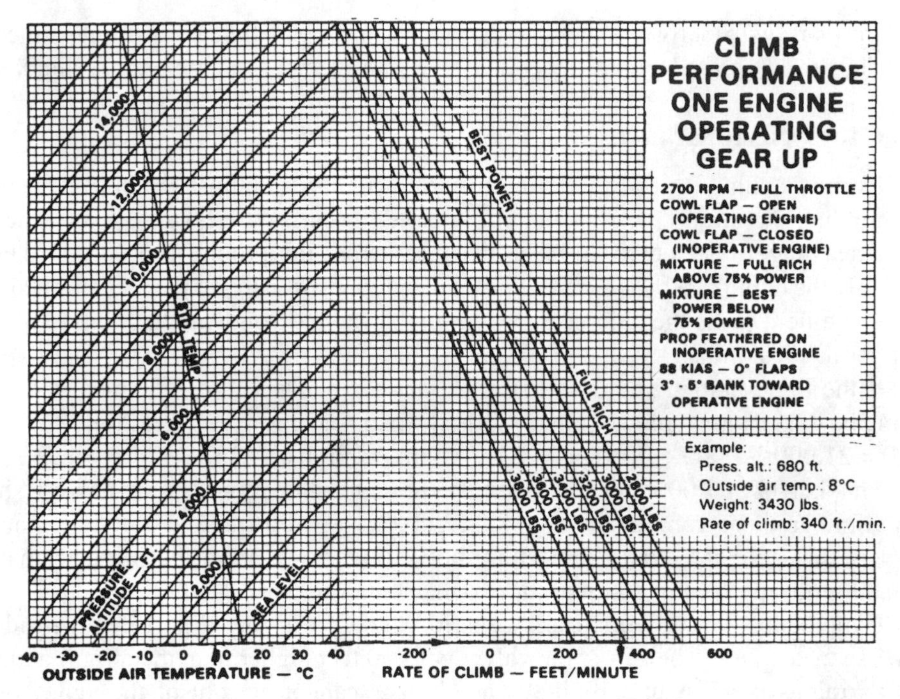

Fig. 2-6. *Climb performance chart: one engine operating. (Copyright Piper Aircraft Corporation. Used with permission.* ***WARNING:*** *This performance chart is presented for illustration purposes only and is not intended to predict the performance of any specific airplane. Piper Aircraft Corporation is not involved with publishing this book and does not endorse any statements made herein.)*

and an airplane weight of 3,430 pounds, the graph says that the airplane will climb at 1,570 feet per minute with both engines operating and landing gear up.

Now look at the same conditions with only one engine. This sample yields a climb rate of only 350 feet per minute. The difference between two-engine climb (1,570 fpm) and single-engine climb (350 fpm) is approximately a 78-percent reduction in performance.

Now look at this single-engine climb performance chart a different way. Read it "backwards." Start at the rate-of-climb scale. Go to the zero-climb position and work back through the chart. If the airplane weighed 3,600 pounds and the temperature was 30°C, the airplane could not climb when starting above a pressure altitude of approximately 4,000 feet. At a maximum weight of 3,800 pounds, the airplane cannot climb on one engine above about 3,000 feet. If this does not scare you, it should, because takeoffs are often made at these pressure-altitude values, even when the airport elevation is near sea level.

If an engine failed just after takeoff (a 3,800-pound airplane and 3,000-foot pressure altitude), the pilot could not climb to safety. It is impossible. Any seconds the pilot uses up trying to convince the airplane to climb is time wasted. Physical science says the plane will not fly to safety, and the pilot should spend these seconds preparing for a landing, even a crash landing, rather than on a vain attempt at an impossible climb.

Other circumstances

Even if a climb is possible, flying to safety might not be possible due to the area surrounding the airport. A 350-fpm climb is very weak. Where will the airplane be upon reaching 1,000 feet AGL? The distance across the ground can be calculated. Multiply the climb rate by 60 and divide by the groundspeed to get the feet climbed per nautical mile. Divide the altitude above ground level that you want to climb to by the feet per nautical mile.

If the light multiengine in the example has a groundspeed of 90 knots, the pilot would be 4.28 nautical miles from the airport when reaching 1,000 feet AGL (350 fpm $\times$ 60 = 21,000; 21,000 $\div$ 90 knots = 233.33 fpm; 1,000 feet $\div$ 233.33 fpm = 4.28 nm).

This height of 1,000 feet and a distance of 4.28 nautical miles forms a very shallow climb gradient from takeoff to 1,000 feet AGL. Any obstruction or terrain that intersects this gradient would force the pilot to maneuver an ailing airplane that is already requiring his full concentration to maintain a shallow climb.

Hypothetically, the worst-case scenario would place the airport down inside a bowl-shaped valley. The rim of the valley is 1,000 feet higher than the airport and the bowl's radius is 4.0 nautical miles. The airplane cannot get out of the valley on a straight-line course; turning the ailing airplane to remain clear of the rising terrain would be difficult.

Worse circumstances

What if the conditions were worse than this? Assume the airplane's weight is 3,800 pounds and the temperature is 30°C, but this time the pressure altitude is 5,000 feet. Working the graph with these numbers yields a "best" performance of –100 fpm. The absolute best climb in this case is a descent. If you were the pilot in this situation and did everything correctly (holding V_{YSE}, 3°–5° bank, gear up, etc.), you would still come back down to the ground.

This presents the pilot with two choices:

- Abort the takeoff at the first sign of engine failure by landing and braking.
- Attempt to fly away but the airplane will settle to the ground anyway.

The airplane is going to land or crash-land either way. The second choice will place the accident farther from the airport, which means farther from flat terrain and rescue crews. Flying away will also tempt the pilot to raise the nose in a hopeless effort to gain altitude; hence, the airspeed will bleed away to V_{MC} and the airplane will go out of control.

A pilot must only accept a climb that is steep enough to provide safety. If the pilot elects to continue an engine-out takeoff and subsequent climb, but the airplane is not climbing fast enough to reach a safe altitude, the accident will just take place farther from the airport.

This adds another item to the decision list:

Speed range	*Pilot action when one engine quits*
V_{YSE} to V_Y	Option 1: Retard throttles, land ahead, start braking. Option 2: Accept single-engine climb rate, climb to safety.

The pilot cannot breathe easy and decide that the takeoff has been successful until an altitude is reached where, if an engine quit, he could maintain altitude clear of any obstacle, or he could descend back to the airport. I affectionately call this V_{BE} for "breathe easy" speed. (V_{BE} is not a recognized V speed in aircraft operation manuals or training syllabuses.) The milestones then are:

Milestone	*Pilot action when one engine quits*
Zero	Tow airplane back to hangar.
V_{MC}	Control but no climb, abort takeoff.
V_R	Liftoff but no climb, abort takeoff.
V_{YSE}	Best single-engine performance, accept or reject climb gradient.
V_{BE}	Safe altitude is available back to airport.

Get into the habit of reading and working the performance charts for your airplane and make the go-no-go decision before takeoff. The decision requires more thought than just the runway length; think about the terrain surrounding the airport. Look for a single-engine escape route that will permit a safe single-engine climb (Fig. 2-7). When a safe altitude is reached, you can afford to return to the airport and land.

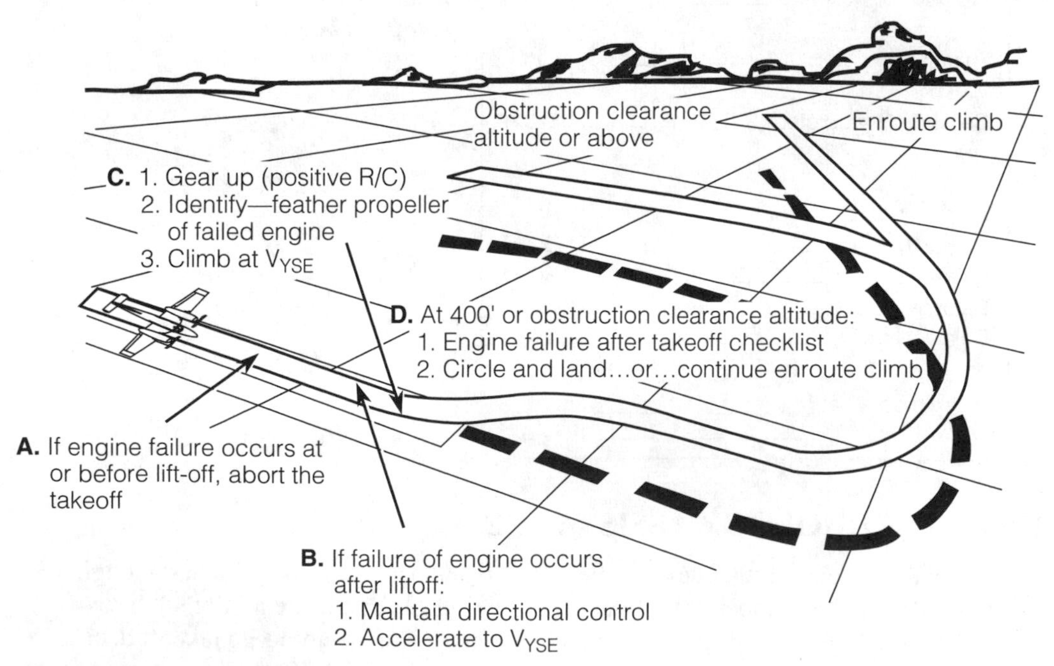

Fig. 2-7. *Flight profile: engine failure on takeoff.*

It is also a good idea to prepare a card that will force you to work the performance problems and help you make an informed takeoff decision. Figure 2-8 is a sample takeoff decision card. Fill in the appropriate data for your airplane and make some copies. Fill in the weather information of pressure altitude, temperature, and wind before every takeoff. Work through the performance graphs and determine the ground roll, accelerate-stop distance, and climb performance. Comparing the crucial information on a one-stop reference card will make the go-no-go decision easier.

The takeoff decision

V_{MC} ________ V_R ________

V_{YSE} ________ V_Y ________

Airport temperature ________ °C

Airport pressure altitude ________ ft

Runway headwind comp ________ kts

Airplane takeoff weight ________ lbs

Runway length (& overrun) ________ ft

Ground roll estimate ________ ft

Accelerate - stop distance ________ ft

Best climb predicted (SE) ________ fpm

Speed milestones

Below V_{MC} of ________ kts ABORT
Below V_R of ________ kts ABORT
Below V_{YSE} of ________ kts ________
Below V_Y of ________ kts ________

"Breath easy" at ________ kts
& ________ ft agl

Fig. 2-8. *The takeoff decision card. (Make some photocopies of this page and put them in your flight bag.)*

HIGH TECHNOLOGY ASSISTANCE

With so many speeds, climb rates, engine settings, and decisions, the pilot has a tough job. Considering the number of high-technology products that make a pilot's job easier, you would think there would be something to help with the go-no-go takeoff decision. There is, the *takeoff performance monitoring system* (TOPMS).

The NASA Langley Research Center in Hampton, Virginia, took up the challenge of providing real-time takeoff information to pilots as a result of the Air Florida crash in Washington, D.C. (The Air Florida 737 crashed after takeoff and hit a bridge before sinking into the Potomac River.) Investigators subsequently determined that the accident was probably caused by improper engine instrument readings as a result of ice buildup. The take off might have been aborted and the accident might have been avoided if the pilots had had supplemental information. NASA set out to provide just such information.

Every spring for the past several years I have taken my flight dynamics class to NASA Langley to visit the wind tunnels. One year the visit included a session with David B. Middleton, NASA aerospace engineer, to discuss the monitoring system. TOPMS is essentially computer software that acts like a stethoscope on the airplane systems. If anything is ever detected that is below performance standards during takeoff, the computer will determine this much faster than the pilot. The system will then warn the pilot of the problem.

The TOPMS system was tested in NASA's flying laboratory, a Boeing 737. The front of the airplane has a standard cockpit and panel (Fig. 2-9). The cabin has an entirely different rear cockpit (Fig. 2-10). The airplane can be flown by remote control from the rear cockpit. Flying from the rear cockpit allows NASA flight crews the opportunity to evaluate new computer hardware and software for future use.

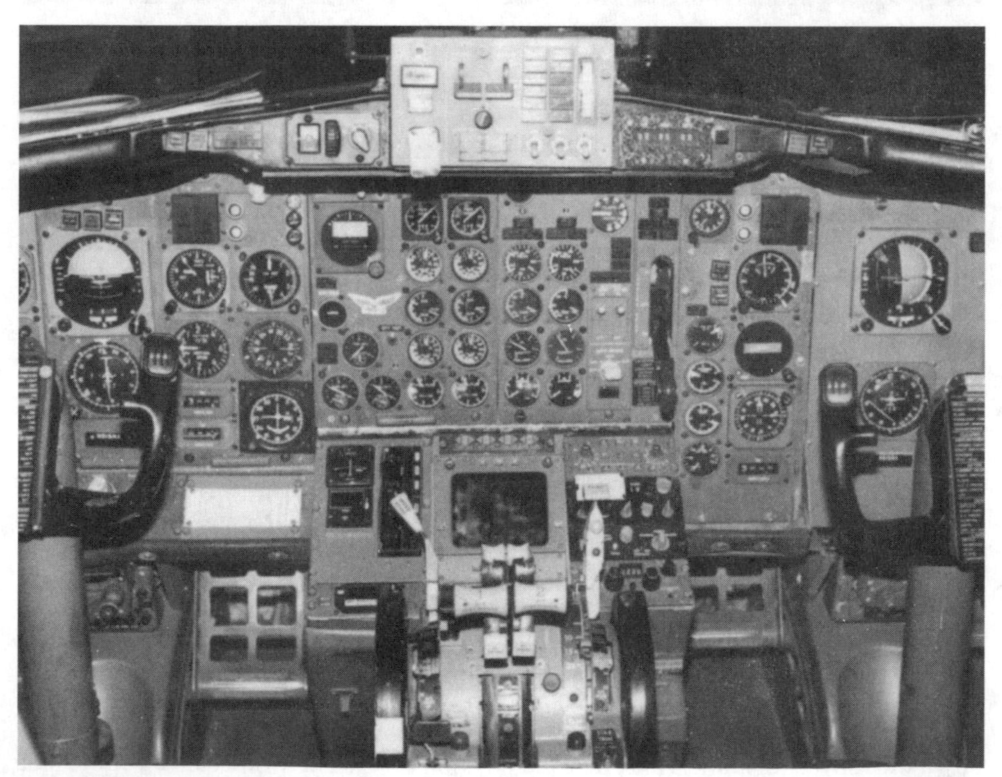

Fig. 2-9. *The forward cockpit of NASA's Boeing 737 has a standard panel.*

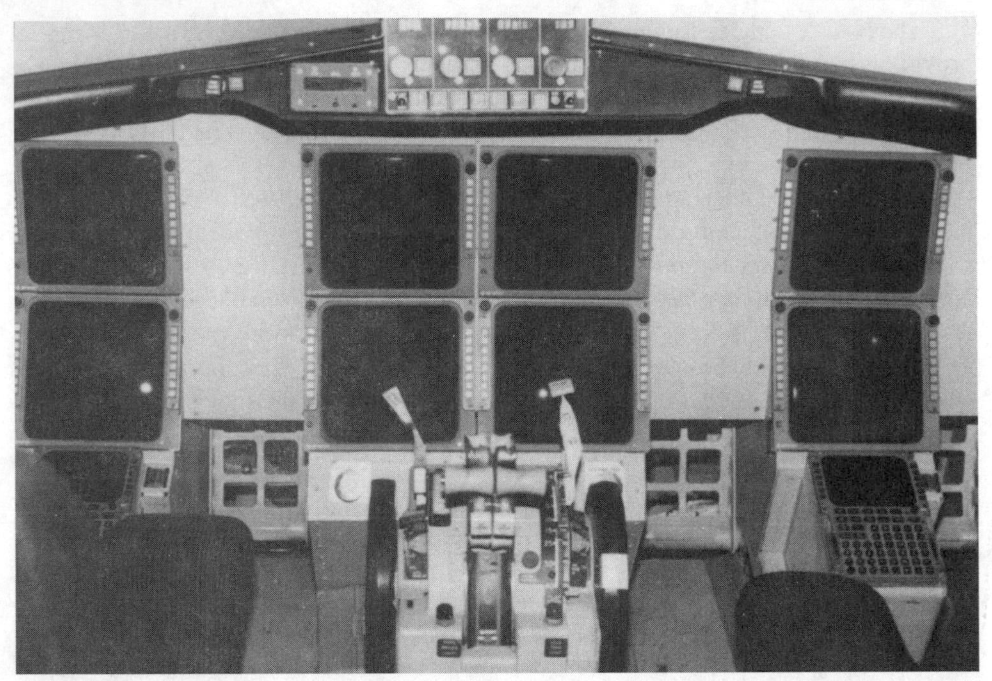

Fig. 2-10. *The rear cockpit of the NASA 737 is a "glass cockpit" where new software, such as TOPMS, is first tested.*

Before takeoff, the TOPMS computer evaluates airplane center of gravity, gross weight, flap settings, ambient temperature, pressure altitude, wind direction, wind velocity, and runway rolling-friction coefficient value. The factors are just like the factors used in the takeoff performance charts. The computer will calculate the expected takeoff performance from these numbers, then determine the distance from the beginning of the takeoff roll to V_R. It also predicts the accelerate-stop distance from zero to V_R and back to zero.

During the takeoff roll, the computer constantly compares the expected performance with the actual performance. The computer evaluates airplane flap setting, left and right throttle positions, left and right engine thrust, calibrated airspeed, airplane acceleration, and ground speed. The pilot is alerted anytime the predicted numbers and actual numbers do not match.

The alert is either an instrument display on the panel, or a head-up display projection on the inside of the pilot's side window. Figure 2-11 is a photo of the panel and head-up display as seen from the pilot's seat. Figure 2-12 shows the head-up display with greater detail.

Computer-aided decisions

The pilot sees an airplane proceeding down a runway. The triangle that is midway down the runway display is the predicted position of V_R. The white bars on either side

Fig. 2-11. *Pilot's TOPMS instrument display.*

Fig. 2-12. *TOPMS head-up display.*

of the runway display indicate the power output of each engine. Both engines are shown here with proper takeoff power. If one engine falters, the power bar of the corresponding engine would become shorter, indicating a percentage reduction in power. An engine failure would cause the power bar to turn red on the side of the engine failure. The number 77 on the left of the display indicates the airplane's current calibrated airspeed.

The pilot would see Fig. 2-13 when the computer has detected a problem that will prevent a safe takeoff. An unmistakable stop sign appears and there is an X on the runway. The X displays where the airplane will come to a stop if the pilot initiates immediate abort action now. This display indicates that the airplane could be stopped safely on the remaining runway.

A TOPMS system has yet to be installed in an air carrier. TOPMS is still several years away from affordability for general aviation. At one time, a global positioning satellite receiver in a general aviation airplane was unheard of. Can TOPMS be very far from reality? (I will trade the GPS for a TOPMS whenever possible because a takeoff monitoring system will make my flight safer than having a satellite navigation system.)

Even if high-technology takeoff monitoring does become available to pilots, every takeoff must be evaluated with respect. Don't ask yourself "What will I do if one engine quits?" Make a declaration: "An engine *will* quit and this is what I will do about it!"

NASA Langley Research Center

Fig. 2-13. *A problem has been detected, and the airplane can be stopped safely.*

TAKEOFF STANDARDS

The following is an edited excerpt from the multiengine practical test standard describing what pilot examiners expect from would-be multiengine pilots on takeoff. The first item on the list refers to the ability to "explain the elements of normal takeoff, including airspeed configurations." This means you will discuss the speed milestones for your particular airplane on the checkride.

Objective. To determine that the applicant:

- Exhibits knowledge by explaining the elements of normal and crosswind takeoffs and climbs including airspeeds configurations and emergency procedures.
- Adjusts the mixture control as recommended for the existing conditions.
- Notes any obstructions or other hazards in the takeoff path and reviews takeoff performance. (Elements of takeoff performance are also required discussion topics.)
- Verifies wind condition.
- Aligns the airplane on the runway centerline.
- Applies aileron deflection in the proper direction as necessary.
- Advances the throttles smoothly and positively to maximum allowable power.

- Checks the engine instruments.
- Maintains positive directional control on the runway centerline.
- Adjusts aileron deflection during acceleration, as necessary.
- Rotates at the airspeed to attain liftoff at V_{MC} plus 5 knots or the recommended liftoff airspeed and establishes wind-drift correction as necessary.
- Accelerates to V_X (best angle of climb airspeed) ±5 knots.
- Retracts the wing flaps as recommended at a safe altitude.
- Retracts the landing gear after a positive rate of climb has been established and a safe landing cannot be accomplished on the remaining runway, or as recommended.
- Climbs at V_Y plus or minus 5 knots to a safe maneuvering altitude.
- Maintains takeoff power to a safe maneuvering altitude and sets desired power.
- Uses noise abatement procedures, as required.
- Establishes and maintains a recommended climb airspeed, within plus or minus 5 knots.
- Maintains a straight track over the extended runway centerline until a turn is required.
- Completes the after-takeoff checklist.

If a crosswind condition does not exist, the applicant's knowledge of the task will be evaluated through oral testing. The examiner will:

- Ask the applicant to explain the elements of normal and crosswind takeoffs and climbs, including airspeeds, configurations, and emergency procedures.
- Ask the applicant to perform normal and crosswind takeoffs and climbs, and determine that the applicant's performance meets the objective.

MAXIMUM PERFORMANCE TAKEOFF

Single-engine pilots are required to do short-field and soft-field takeoffs, which fall into a maximum-performance category. Multiengine pilots are also faced with takeoffs requiring the airplane and the pilot to perform their best. The maximum-performance takeoff in a multiengine airplane is not subdivided to short- and soft-field. A soft-field technique can be practiced by holding more elevator in the takeoff roll, which reduces the load on the nosewheel, but the airplane should never be allowed to leave the ground with the airspeed at or below V_{MC}.

The maximum performance takeoff requires the pilot to be in even greater control of airspeed because the airplane must be placed in a climb that is steeper and slower that normal. If the climbout is slower, it is also closer to V_{MC}. After V_R and liftoff, the nose of the airplane is raised to achieve the best angle of climb, V_X. This climb angle places the airplane at the highest possible altitude in the shortest distance across the ground to clear an obstruction at the end of the runway.

Normal takeoff and climb can be divided into three segments:

- Takeoff roll from zero to V_R.
- Climb at V_Y until a safe altitude is reached.
- Adjust power and pitch to achieve a cruise climb and best cooling airflow through the engines.

A maximum-performance takeoff adds one more segment that includes another speed milestone:

- The takeoff roll accelerates from zero through V_{MC} to V_R, which is at least V_{MC} plus 5 knots.
- The airplane leaves the ground and continues to accelerate, but the pilot raises the nose higher than normal and stops the acceleration at V_X. The speed of V_X is held steady by the pilot until all obstructions are clear.
- The pilot allows the nose of the airplane to fall and the airspeed to accelerate to V_Y until a higher, safer altitude is reached.
- A cruise climb is established.

Standards

The following edited version of the FAA's Multiengine Practical Test Standards regards maximum performance takeoffs. The examiner will ask the applicant to explain the elements of a maximum performance takeoff and climb, including the significance of airspeeds and configurations, emergency procedures, and the expected performance. The examiner will also ask the applicant to perform a maximum performance takeoff and climb, and determine that the applicant's performance meets the objective. The examiner's objective is to determine that the applicant:

- Exhibits knowledge by explaining the elements of a maximum performance takeoff and climb, including airspeeds, configurations, and expected performance for specified operating conditions.
- Selects the recommended wing flap setting.
- Adjusts the mixture controls as recommended for the existing conditions.
- Positions the airplane for maximum runway availability and aligns it with the runway centerline.
- Advances throttles smoothly and positively to maximum allowable power.
- Checks engine instruments.
- Adjusts the pitch attitude to attain maximum rate of acceleration.
- Maintains positive directional control on the runway centerline.
- Rotates at the airspeed to attain liftoff at V_{MC} +5 knots, or at the recommended airspeed, whichever is greater. (The applicant explains the extra takeoff segment during a maximum performance takeoff.)

- Climbs at V_X, ±5 knots, or the recommended airspeed, whichever is greater until obstacle is cleared, or to at least 50 feet above the surface, then accelerates to V_Y, +5 knots. (The applicant explains the extra takeoff segment during a maximum performance takeoff.)
- Retracts the wing flaps as recommended at a safe altitude.
- Retracts the landing gear after a positive rate of climb has been established and a safe landing cannot be made on the remaining runway, or as recommended.
- Climbs at V_Y, ±5 knots, to a safe maneuvering altitude.
- Maintains takeoff power to a safe maneuvering altitude and sets desired power.
- Uses noise-abatement procedures as required.
- Establishes and maintains a recommended climb airspeed, ±5 knots.
- Maintains a straight track over the extended runway centerline until a turn is required.
- Completes the after-takeoff checklist.

After the initial climbout has been completed, most airplane manufacturers recommend that the pilot "clean up" the airplane. Most light twins use full power during takeoff, but full power is not required during the entire climb to altitude. It is a good idea to establish a power-reduction routine that is followed for after every takeoff. I call for a "500-foot check," which means that I am at least 500 feet above the ground with no obstructions.

The power can be reduced from full takeoff power to a cruise climb power setting. The manifold pressure is usually reduced to 25". Be careful. Most multiengine students mistakenly look at the tachometer; make absolutely certain that you are looking at the manifold pressure gauges. Propellers can be adjusted for an efficient climb after the power is properly set.

The RPMs are usually brought back from full forward to 2,500 RPM. This 25"/2,500 RPM combination is sometimes referred to as "squaring off" and can just be called 25/25. The nose should also be lowered and airspeed allowed to accelerate from V_Y to a "best cooling" climb speed. Check your airplane's handbook for this speed. The cruise climb can continue with the engine and propellers at 25/25 and good airflow through the engine compartments.

Never let your guard down when it comes to takeoff. Study the existing conditions and predict the airplane's performance. Decide *before* you taxi onto the runway what you will do when an engine quits. Make the takeoff in your mind and simulate engine failures at every speed. By doing this, your actions during an emergency will be swift and undelayed by indecision. You have a better chance of making the correct decision during the calm of taxi when a takeoff is still an option, rather than in the heat of battle on the runway.

MULTIENGINE LANDINGS

The multiengine airplane employs the same aerodynamics for landing as a single-engine airplane. The major difference is that approach speed is faster. Good planning for a short runway is essential so that excessive braking or an unnecessary go-around can be avoided.

The pilot must first get accustomed to slowing the airplane down during traffic pattern entry or at some position during a straight-in approach. Then airspeed control must be maintained throughout the pattern to final approach. Approach speed should be V_{YSE} or the blue line until over the runway and at the beginning of the flare for touchdown. This lengthens the landing, but it will pay off during a go-around.

An engine failure in flight will make you very anxious to find a suitable runway and get on the ground. There is some good news. Landing with one engine failed is no more of a problem that a normal landing. The correct normal landing procedure involves gradual power reductions closer to touchdown, eventually reducing the power to idle when the landing is made.

The yaw tendency is reduced every time power is reduced. Rudder correction diminishes through touchdown, when it will probably be unnecessary. Stay well above V_{MC} until over the runway and concentrate on a normal touchdown.

Normal, crosswind, and single-engine landings, and go-arounds, are discussed in detail in chapter 7, regarding multiengine flight training.

3
Engine-out procedures

WHEN AN ENGINE DOES FAIL, THE PILOT MUST REACT CORRECTLY AND without hesitation. Having two engines is an advantage, but the pilot must place the airplane in a position to utilize the advantage. The first reaction from the pilot depends on the airplane's speed when the engine fails. If the airplane's speed is slower than V_{MC}, the pilot must reduce the power on both engines. If the speed is faster than V_{MC}, the pilot must increase power on both engines. Making the wrong choice here is fatal.

If one engine fails, the pilot will feel the sway-yaw of the airplane. It can happen fast, and in that first second before you realize what is happening, it can surprise even the most veteran multiengine pilot. In that second, it might be unclear which engine has failed because the airplane is reacting to the failure faster than the pilot can recognize it.

Because the exact engine's failure has not yet been positively determined, the action taken by the pilot should include both throttles. By moving both throttles in unison, the pilot is sure to affect the operating engine, even if the pilot does not know which one is operating during this moment of confusion. It is better to move both throttles even though one is dead than to waste time figuring out which one works and which one does not work.

PROPER THROTTLE ADJUSTMENT

Now we understand that both throttles should move together, but which way should they move? If the airplane is slower than V_{MC}, both throttles should be pulled back. If the pilot pushes the throttles to a higher power setting, directional control will be lost. If both throttles are brought forward, the dead engine will continue to produce no power, but the operating engine at full power will produce the yaw that the rudder cannot overcome while slower than V_{MC}.

This idea of "pull back when in trouble" goes against our earlier training. In single-engine airplanes, the remedy for a stall was to push forward on the wheel to break the stall and push forward on the throttle. We associate recovery with full power. Some of us have become "Pavlov's pilots" because we hear a horn and add power without thinking.

Below V_{MC}, power must be reduced on both throttles; the power output from each engine will be the same: zero. Having zero power is not good, but at least it is equal power and no yaw will result. With the power pulled back on both engines, lower the nose and gain airspeed.

When the airplane is traveling faster than V_{MC}, the pilot can afford to add power on the good engine while opposing yaw with rudder. This requires the airplane to have sufficient altitude to gain the speed. If there is no altitude, the pilot should land straight ahead on the flattest, softest thing available.

If the airplane's speed is faster than V_{MC} when an engine failure is felt by the pilot, both throttles should go forward. The greatest power will set up the best possible climb performance for the situation. Unfortunately, the pilot cannot throw both throttles forward if the propeller controls are back. If the engine failure occurs during cruise flight, the props will be back.

Asking the propeller in a high-pitch/low-RPM setting to take on full engine power is like putting the force of a cannon through the barrel of a rifle. It is not matched and can damage the propeller and engine. This is why it is the proper procedure to place the prop controls full forward for takeoff and for landing.

In this position, if full power is required, the props are already prepared to handle the extra load. Most manufacturers also recommend that the engine mixtures be at full rich because a faulty mixture might cause an engine to fail. Consult your own airplane's manual, but usually the proper response to a faster-than-V_{MC} engine failure is:

1. Both mixtures forward (rich).
2. Both propeller controls forward (high RPM).
3. Both throttles forward (high power).

These steps should be taken when the pilot feels the yaw of an engine failure. At the same time, the pilot must "fly the airplane." If the airplane's speed is faster than V_{MC} at the time of the failure, the pilot cannot allow depletion of the airspeed to V_{MC} while working in the cockpit to understand and fix the problem.

MAINTAIN AIRSPEED

When the yaw of engine failure is first felt, add rudder to keep the airplane straight, adjust the airplane's pitch as necessary to maintain a speed above V_{MC}, then go to work on the problem. Many pilots have panicked here, and while they were moving levers and throwing switches, the airplanes sank below control speed and crashed.

The safe multiengine pilot must have a little savvy. The safe multiengine pilot takes calculated time to fly the airplane and then go to work on the airplane. The mental checklist would read:

1. Apply rudder as the unknown yaw sets in.
2. Control airspeed above V_{MC}.
3. Both mixtures to full rich.
4. Control airspeed above V_{MC}.
5. Both prop controls to high RPM.
6. Control airspeed above V_{MC}.
7. Both throttles to full power.
8. Control airspeed above V_{MC}.
9. Use additional rudder as the power comes up on the good engine.
10. Maintain airspeed above V_{MC}.

This all needs to happen fast, but not in a mindless blur. Multiengine pilots must be alert because the thought process can be challenged at any moment. When the initial shock and reaction has passed, and the airplane is under control, it will be time to think about performance.

Recall from chapter 2 that single-engine climb performance is bad at best. In order to give the airplane the best chance of staying in the air, the pilot must now reduce drag so the remaining good engine can do its job.

The wing flaps are also drag producers, and they should be retracted to achieve the best possible climb unless the manufacturer has prescribed a "best-climb flap setting." If the airplane is "stabilized" in level flight or in a climb, the pilot should also bring the landing gear up.

It is vital that the decision to continue flight, as opposed to landing on the remaining runway (if any), is made before gear retraction. The thinking here is that if you are going down anyway, you should have the landing gear out to help absorb the shock of impact. But if a climb is possible, the landing gear sticking out into the wind is a tremendous drawback to performance and the pilot is better off with gear up.

The thought progression to take care of the power is: mixture, props, and throttles. Also verify that the fuel pumps are on, then reduce drag: flaps up and landing gear up.

A pilot must set priorities to manage this engine-out crisis. Each move the pilot makes is calculated to yield the best airplane result at the best time:

1. Understand that an engine has failed.
2. Counteract yaw with rudder.
3. Control airspeed.
4. Mixtures rich.
5. Props high RPM.
6. Throttles to full power.
7. Wing flaps retracted or to recommended setting.
8. Landing gear up when level or climbing.

The first three items assure airplane control. The next five items work toward giving the airplane the best performance. Now it is time to find out what has happened to the airplane.

AIRBORNE DETECTIVE

The failed engine must be identified before any further corrective action can be taken. Remember that no instrument on the panel clearly spells out which engine has failed. The failed engine will still be turning because it is being driven by the wind; therefore, the tachometer will show nearly the same RPM on both engines, even though one is not producing power.

When the propeller turns, the pistons are still moving up and down in the cylinders and still drawing in air. For this reason the manifold pressure gauge is also reading near normal. The windmilling propeller on the dead engine is still turning fast enough so the blades are blurred to invisible. This means that the pilot cannot tell which engine is the problem by just looking at it, unless smoke or oil is evident.

The dead engine is identified by "feeling" the yaw produced by the good engine. If the dead engine is on the right side, the airplane will yaw to the right. This will require the pilot to push the left rudder pedal. With left rudder applied, the nose (speeds above V_{MC}) can be returned to a straight position.

Now the pilot must hold the left rudder continuously in opposition to the good engine's yaw. At this time, there is no need for any right rudder, so the pilot's right foot is not required. This leads to the saying "Idle foot/idle engine," meaning that the side where no rudder is needed is also the side of the failed engine. Determining which engine is inoperative is accomplished with the feet.

When the failed engine is identified, the pilot should verify that the correct decision has been made. This is done by pulling back the throttle on the engine that is believed to have failed. If the pilot has made the correct decision, nothing will happen because the engine has quit anyway, and the throttle position does not matter.

If the wrong decision is made and the good engine's throttle is pulled back, the pilot will instantly feel the difference. The good-engine yaw will go away, and the cockpit will be much quieter. If you do pull back on the wrong engine throttle, reapply full power.

This is where the engine-out checklist has evolved so far:

1. Understand that an engine has failed.
2. Counteract yaw with rudder.
3. Control airspeed.
4. Mixtures rich.
5. Props high RPM.
6. Throttles to full power.
7. Wing flaps retract or to recommended setting.
8. Landing gear up when level or climbing.
9. Identify failed engine (idle foot/idle engine).
10. Verify failed engine (reduce failed engine's throttle).

Perform the checklist as soon as a faster-than-V_{MC} engine failure is recognized. Items 1–3 ensure airplane control. Items 4–8 give the airplane the best possible performance. Items 9–10 determine which engine has the problem. But after item 10, the pilot reaches a crossroad.

FIX OR FEATHER?

Fix means the pilot attempts to get the engine restarted; feather means that the pilot either aborts an engine restart or has no time to consider a restart. The determining factors in the decision to fix or feather are airplane speed, airplane altitude, time available, and the reason the engine failed.

If the airplane is high and fast at the time an engine fails, the pilot might pause after item 10 on the mental checklist and attempt to discover the problem. Height and speed give the pilot the luxury of time. The reason the engine has quit could be *relatively* minor. Perhaps improper fuel management has caused a tank to run dry, and subsequently an engine quits; the pilot could switch the fuel valve to a tank that still contained fuel and restart the engine.

But at low altitudes and slow airspeeds, the pilot is robbed of time to fix the problem. Even if the engine failure were fixable, the pilot could not spend time attempting a fix during a takeoff or initial climbout. In those critical situations, the best thing to do is reduce drag by feathering the prop.

If the pilot determines that it is safe to attempt a restart, the following troubleshooting checklist might be helpful. Always consult the airplane's operating handbook for specific instructions.

TROUBLESHOOTING

1. Complete procedures for engine failure (items 1 through 10).
2. Decide to fix or feather. If the decision is to fix, do not feather and initiate troubleshooting.
3. Troubleshoot. (*Continued on page 62.*)

a. Fuel selectors ON.
b. Primers LOCKED.
c. Carburetor heat ON.
d. Mixture ADJUST.
e. Fuel quantity CHECK.
f. Fuel pressure CHECK.
g. Oil pressure and temperature CHECK.
h. Magnetos CHECK.
i. Fuel pumps ON.

An improperly set fuel selector, carburetor ice, improper mixture, or low fuel pressure can cause an engine to quit. This troubleshooting checklist covers all the *minor* problems that could arise and be corrected to get the engine up and running again; perhaps your knee inadvertently hit the magneto switch. You cannot be a mechanic in the air, but if time and altitude permits, you should check what is possible. If, after these troubleshooting items are complete, you realize that the engine cannot be restarted, then feather the prop and secure the engine (described in the next subsection).

If the engine failure occurs on takeoff or during the initial takeoff climb, there will be no time for the troubleshooting checklist. After the failed engine has been properly identified and verified, feather the propeller using the manufacturer's guidelines. Usually this means pulling the prop control all the way back into the feather position.

Most multiengine prop controls have a detent in the travel of the lever. The detent is a place where the lever rubs in its track. If a pilot pulled back on the prop control inadvertently, she would feel this resistance on the lever before going as far as the feather position. Theoretically, a propeller can only be feathered on purpose.

Always go through the identify-and-verify process before feathering the propeller. Just about the worst thing you can do (besides landing gear up) is to feather the good engine. When the decision is made to feather, follow the feather procedure provided in the airplane's manual.

Secure the engine

After the propeller has been feathered, the job is still not done. If the airplane is up high, the pilot should secure the engine. If the airplane is down low, and the pilot is struggling to maintain control and climb, then the last thing on the pilot's mind will be another checklist. Use good judgment here. Problems might worsen without securing the engine. Follow the manufacturer's recommendations to secure an engine. This checklist is an example:

1. Mixture IDLE CUT OFF.
2. Fuel selector OFF.
3. Fuel pump OFF.
4. Magnetos OFF.
5. Alternator OFF.

6. Cowl flap CLOSED.
7. Reduce electrical load as required.
8. Maintain at or above V_{YSE}.

Without such a list, fuel would continue to be pumped to a hot, disabled engine, which might cause a fire. You have enough problems without also starting a fire. After the engine is secure, the pilot can turn attention to safely getting back on the ground. Reduce power on the operating engine, if that is possible, to protect that engine from overheating and wear due to extended use at a power setting that is higher than normal.

ENGINE FAILURE SUMMARY

When an engine quits, the pilot's mind cannot. The engine failure events look like this:

1. Engine quits.
2. Pilot recognizes that an engine has quit.
3. Pilot controls airplane.
4. Pilot gets best performance from airplane.
5. Pilot decides to fix or feather.

6A. Pilot feathers prop using proper procedure. (Go to step 7.)

6B1. Pilot troubleshoots the problem, and the engine restarts. (Pilot continues flying, but monitors the situation closely and considers a precautionary landing to determine what happened.)

6B2. Pilot troubleshoots the problem, and the engine fails to restart. (Go to step 6A.)

7. Pilot secures engine.
8. Pilot lands safely.

Each step requires the pilot to fly skillfully and make decisions correctly. All these events can happen quickly. Items 1 through 7 above might take place in less than a minute with no warning. Multiengine pilots must be able to handle emergencies with swift, accurate, and knowledgeable action. Multiengine pilots are only safe if they are competent disaster managers.

4
Multiengine propeller systems

THE PROPELLERS AND PROPELLER SYSTEMS OF A MULTIENGINE AIRPLANE must be capable of doing more than a single-engine propeller. The biggest difference is that a propeller must be stoppable on a multiengine airplane. The value of a featherable propeller was proven in chapter 1 on multiengine aerodynamics. The speed of V_{MC} skyrockets when a prop is windmilling, versus when a prop is feathered. Multiengine airplanes must come with a propeller system that will allow for this drag reduction, and V_{MC} reduction, to occur.

To become a commercial pilot, an applicant must have a minimum of 10 hours training in an airplane that has retractable landing gear, retractable flaps, and a constant-speed propeller system. If you train for the commercial pilot certificate in a single-engine airplane, you will become familiar with a propeller system that changes blade angles, but does not feather. For building-block purposes, it is usually best to progress from a fixed-pitch propeller airplane to a constant-speed propeller, to a constant-speed-full-feathering propeller.

ANGLED FOR EFFICIENCY

It is important to understand why propeller blade angles need to change. Fixed-pitch propellers are manufactured to maintain only one angle, but no single angle will provide the best efficiency all the time. A fixed-pitch propeller is like having only first gear in a car. If the only gear you have is first gear, then everything is great when pulling away from a stop sign, but when you get to about 15 miles per hour, first gear becomes very inefficient. First gear at speeds above 15 mph will strain the engine and burn extra gas.

Fixed-pitch props are usually set with blade angles efficient for either climb or cruise. Flying a single-engine airplane equipped with a fixed-pitch climb propeller is like being stuck in first gear. You get off the runway just fine, but when you level off for cruise, the prop will hold the airplane back, just as first gear prevents the car from going faster.

Most flight schools own single-engine airplanes with the props in a fixed-pitch climb angle because student pilots need the best takeoff and climb performance to remain safe while learning.

Changing blade angles allows pilots to shift gears as if in a car. At a car speed faster than 15 mph, the driver should shift into second gear, which makes the engine and drivetrain operate more efficiently. A controllable propeller allows the pilot to shift gears. Several systems were designed to accomplish this task. The most common is the constant-speed propeller.

PROPELLER DYNAMICS

What causes propeller efficiency? The propeller blade is simply an airfoil that swings around to create relative wind. Each blade is twisted. Each blade's angle of attack is greatest near the spinner; the angle of attack is gradually reduced moving toward the tip. The blade tip is moving faster than the inner parts of the blade, so the tip requires less angle of attack to produce thrust.

If the propeller blades did not twist, but had a constant angle of attack throughout the length of the blade, then the tips, because of their greater speed, would produce the greatest thrust (Fig. 4-1). But that would cause a problem because the tips are thin and weak. The thrust, if allowed to be greatest at the tips, would bend the tips forward, and this could damage or destroy the propeller.

Fixed-pitch propellers actually have many blade pitches when you consider the different locations along the blade (Fig. 4-2). Now consider just one location on any propeller. Let's look in detail at what is happening at a position 6 inches from the tip, toward the spinner. This position has a particular blade angle, and this angle will always be the same on a fixed-pitch propeller.

Relative windmill

As the blade moves, the air that passes around the blade is the blade's relative wind. Just like the relative wind on the airplane's wing, this relative wind is always

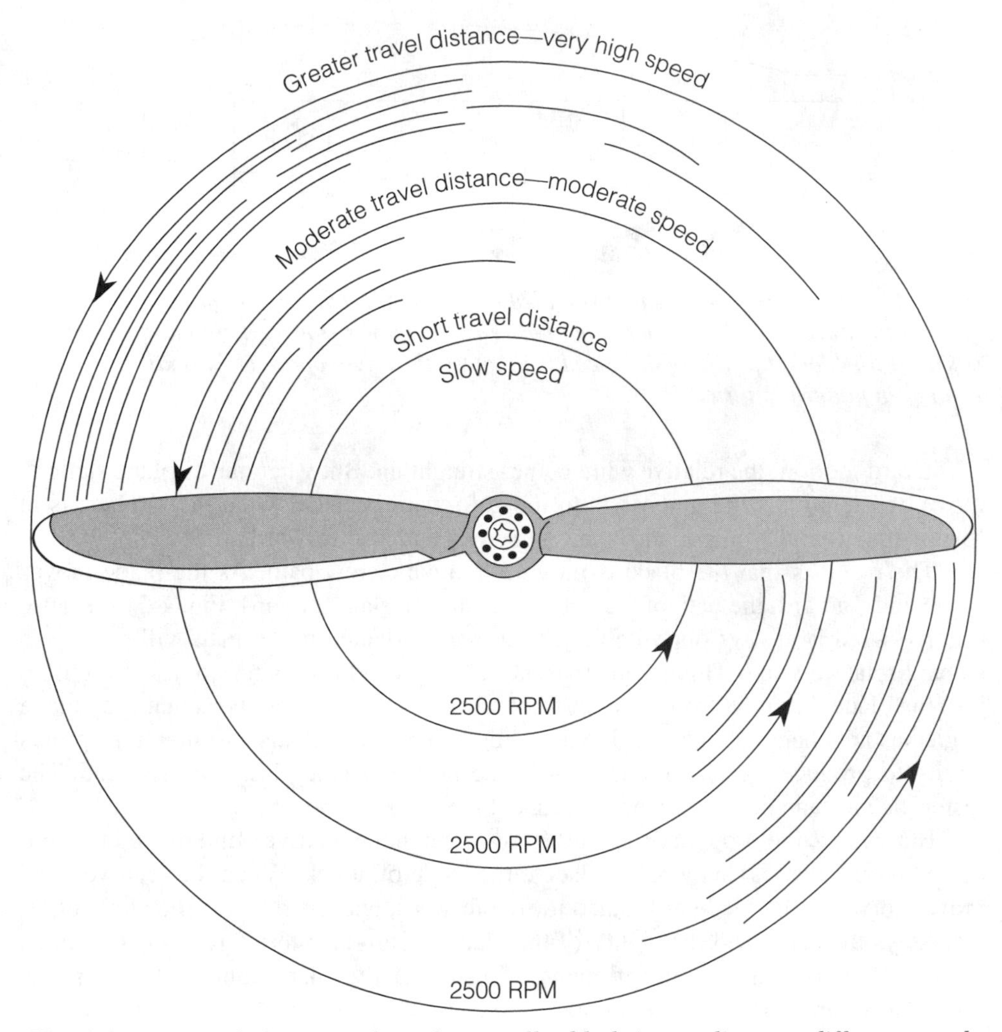

Fig. 4-1. *Every airfoil section along the propeller blade is traveling at a different speed. This requires the blade to "twist" in order to produce an even thrust pattern.*

moving the opposite direction to the blade's travel path. When the prop blade is moving down (the right side of the propeller arc as seen from the pilot's seat), the relative wind at that position is up. When the prop is moving up (left side of propeller arc), the relative wind is moving down. This means that a fixed-pitch propeller is most efficient when the airplane is not in motion! Think about it.

The greatest thrust is produced when the angle of attack is greatest per given RPM. When the airplane's engine is running, but the airplane is not in forward motion, the angle of attack will be the greatest. Here is why: Angle of attack is determined by the angle between the airfoil's chord line and the relative wind. As the prop blade goes down, the relative wind comes up, and this forms the angle. When the airplane is not

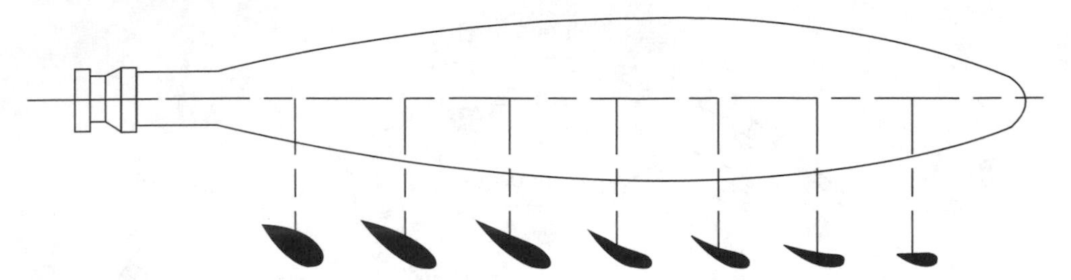

Fig. 4-2. *Each airfoil cross section is called an element. Each element has a different angle of attack. Near the hub, where the blade element moves slowly, a high angle of attack is set. At the blade tip, where the speed is greatest, the angle of attack is small. The result is an even pattern of thrust.*

in forward motion, the relative wind comes straight up. But when the airplane starts to move, the relative wind also starts to move forward because when the airplane is in forward motion, the prop blade is moving down and also forward.

The result is that the blade is moving in a corkscrew path. As the blade moves down and forward, the path of the blade is actually a slant forward (Fig. 4-3). Because relative wind is always opposite the travel path, a slanted travel path will produce a slanted relative wind. This means that the relative wind no longer comes straight up into the blade, but now from more in front of the blade. Because the actual prop blade angle has not changed, this cuts down the angle of attack and this cuts down the prop's ability to produce thrust. In other words, the prop efficiency has been reduced. The faster the airplane goes, the more inefficient the propeller becomes.

But what if the prop blade moved forward when the relative wind moved forward on the prop blades? This would preserve the angle of attack. When the relative wind moves forward, the blade angle also moves forward, and the result is the angle of attack stays the same and the ability of the blades to produce thrust remains the same. This would maintain propeller efficiency. This is exactly what a controllable propeller does. The entire blade twists in its socket to chase the shifting relative wind. In effect, the prop blade shifts gears to remain efficient.

Taking off

For takeoff, the blade angle should be set to a low angle of attack. Because the airplane will have a slow forward speed during takeoff, the propeller blade's angle of attack will be efficient. But a low angle of attack on the blades also reduces drag and allows the engine to turn as fast as possible. Just like any airfoil, the production of lift produces induced drag.

When the drag on the prop is high (large angle of attack), the engine will labor under this additional strain and the RPMs will be reduced. When the drag on the prop is low (small angle of attack), the engine has less to hold it back and the RPMs increase. A low-pitch/high-RPM setting is best for takeoff because this allows more total thrust.

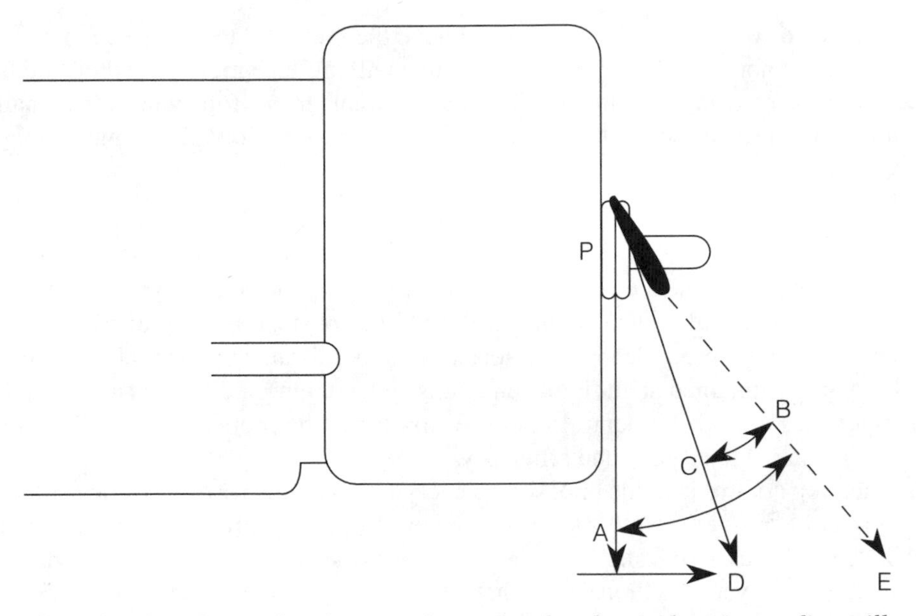

Fig. 4-3. *When the airplane's engine is running, but the airplane is standing still, the propeller blade moves from position P to A. Relative wind is always opposite the direction of travel, so relative wind is from position A to P. The chord line of the blade's airfoil is the extended line P to E. The relative wind and chord line make up the angle of attack, which is angle A to B. When the airplane begins to move forward, the blade moves both down and forward in the direction from P to D. This changes the relative wind to the direction D to P. If the propeller blade does not change its chord line position, the angle of attack is cut down to angle C to B. This lowers efficiency because the "bite" of the air is actually less. The only remedy is to move the blade's chord line out, which is what a constant-speed propeller does.*

Each time the prop rotates, the blades will take a smaller bite of the air, but because this setting produces less drag, the prop can turn faster and take more bites than would be the case with a higher pitch setting.

This idea of blade angles and RPMs is hard to visualize because you cannot actually see the blade angle while the blades are in motion. To help understand why low pitch equals high RPM and high pitch equals low RPM, substitute low angle with low drag and high pitch with high drag.

Use the example of a lawn mower. Will the engine run faster in high grass or low grass? The answer is low grass because the engine has less to impede its blade motion. Low grass produces low drag and high RPM. A thick, high lawn produces more drag, and the engine is burdened and runs slower.

I often ask students this question: "Why don't we take off with a low-RPM and high-pitch blade setting? That way each swing of the prop would take a bigger bite of the air and we could get in the air faster." I hope they respond to this question by saying something about how the higher blade angle will produce more prop drag and that means less prop swings per minute. The smaller number of prop swings, the smaller the total thrust. Even though the bite of air is smaller with a low blade angle, you get more bites.

HARDWARE

This is the physical science reason why a movable propeller blade system is needed, but how does it actually work on the airplane? The constant-speed propeller systems typically move the prop blades to various angles in unison. The most efficient blade angle can be maintained at all flight conditions and without much input from the pilot. The pilot uses a propeller control in the cockpit to set the proper speed, and then the system maintains that speed and efficiency.

The prop control is to the pilot what a stick shift is to the car driver. The car driver might only have five forward speeds, but the pilot's prop control has an infinite number of "gears" to choose from. After the pilot selects the proper RPM for the flight condition, the prop system will maintain that RPM setting by automatically changing the blade angles.

When the RPMs get faster than the prop setting, the prop system will raise the prop drag by increasing the blade angle, and the RPMs will slow down to the prop setting. When the RPMs get slower than the prop setting, the system will reduce the blade angle and RPMs will speed up to the prop setting.

The system is based on a unique balancing act. Systems differ only slightly from manufacturer to manufacturer. Most systems start off with the ability of the blades to turn. Constant-speed propeller blades that turn to various angles are cylinder-shaped where the blade enters the spinner (Fig. 4-4). Fixed-pitch blades usually still have an airfoil shape at the spinner (Fig. 4-5).

The spinner of a fixed-pitch propeller is just for window dressing and drag reduction. The spinner of a constant-speed system is usually fatter because the spinner contains either a large spring, an air pressure chamber, or both. The spring is called the *hub spring* (Fig. 4-6), and left alone, it will push the prop mechanism. The prop mechanism is a piston that moves either forward or aft inside the spinner. The piston is connected to the prop blades so that when the piston moves, the blades turn to a different angle of attack.

In many systems, the hub spring will push the piston aft, which will cause the prop blades to go to a high-pitch/low-RPM position. In order for the blades to return to a low-pitch/high-RPM position, the piston must move forward again in opposition to the force of the hub spring. This opposition force is supplied by engine oil.

Engine oil from the oil pan is allowed to pass into a *prop governor pump* that raises the oil pressure high enough to push the hub spring forward. High-pressure oil is lead to the aft side of the spinner's piston. If the oil pressure is stronger than the hub spring pressure, then the piston will move forward, causing the prop blades to move to a low-pitch/high-RPM position. The balancing act is between oil on the back side and a spring on the front side of the piston.

Fig. 4-4. *This airplane has a constant-speed propeller, which can be easily identified by the shape of the propeller as it enters the spinner. This blade shank has a cylindrical shape that allows the blade to turn in its socket.*

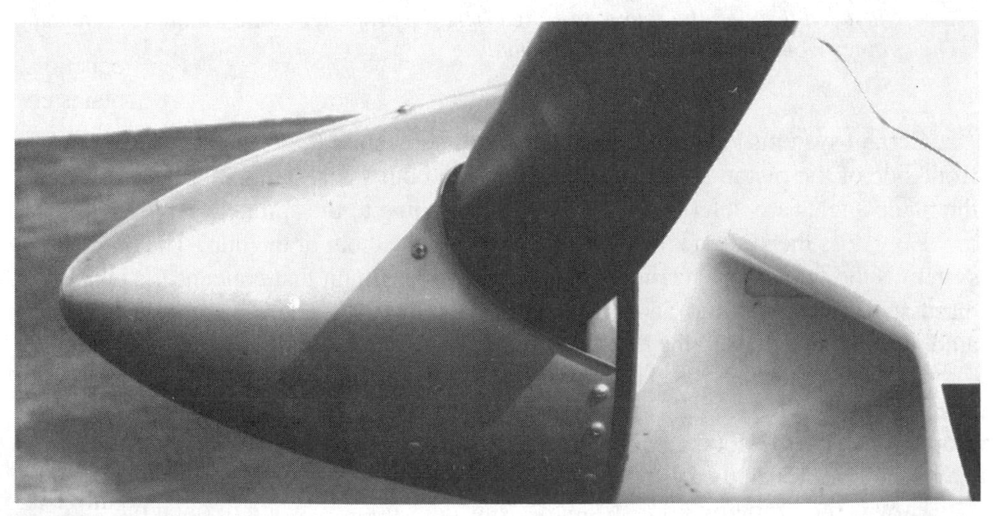

Fig. 4-5. *This airplane has a fixed-pitch propeller. The blade's airfoil shape is intact as the blade enters the spinner. This blade cannot twist in the socket.*

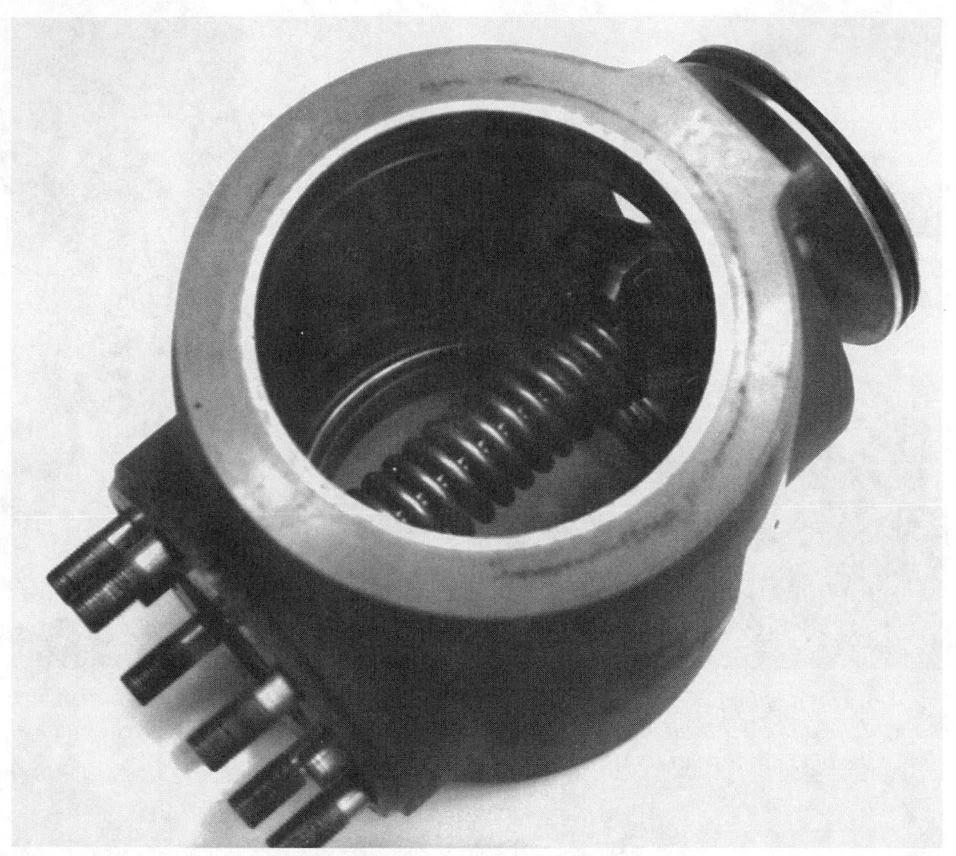

Fig. 4-6. *This is the large hub spring of a constant-speed propeller. The photo of the spring is taken through the round opening where the propeller blade would be attached. The spinner would cover this entire assembly.*

Certain systems balance oil against air pressure, and other systems route oil to the front side of the piston so that oil pressure is on both sides. No matter how it is done, the blade angles are determined by oil pressure going to the spinner.

How does the system know when to send oil in and out of the hub? The brain of the system is the *propeller governor* (Fig. 4-7), a combined unit that contains the previously discussed governor pump and a speed-sensing oil valve. The governor regulates the amount of oil that goes to the hub and therefore regulates the propeller RPM setting.

Whenever the water level in a lake behind a dam gets too high, the dam operator can open up the flood gates, allowing water to flow through the dam and downstream. The prop governor is the flood gate operator. When the system requires more oil, the governor opens the gate. When the system requires less oil, the gate is closed by the governor.

How is the prop governor so smart? How does it know when to open the oil gates and when to close them? The prop governor relies on a delicate balancing act to operate. The opposing forces are the force of a small spring and centrifugal force. The pro-

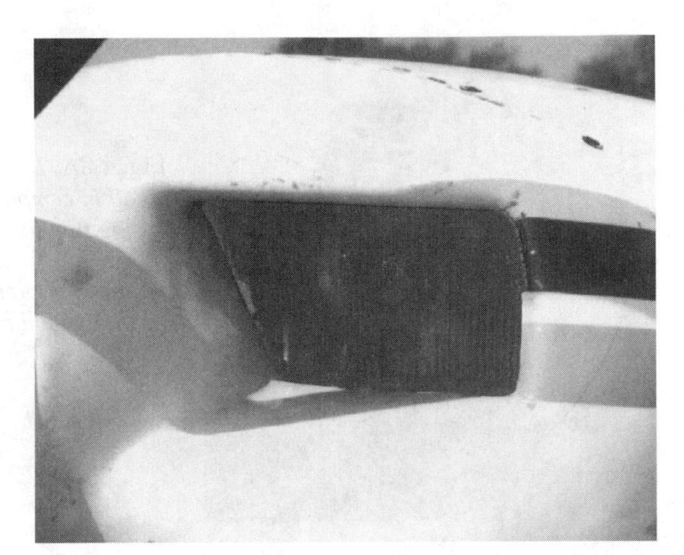

Fig. 4-7. *The propeller governor is often hidden inside the engine nacelle.*

peller governor has a smaller spring inside called the *speeder spring*. Never confuse the larger hub spring, which is inside the spinner, with the smaller speeder spring, which is inside the governor.

When the pilot moves the prop control inside the cockpit, the speeder spring (Fig. 4-8A) is either tightened or loosened; the pilot sets the force of the spring. The spring rides atop a spinning shaft that is connected by gears to the engine. The spinning shaft will speed up or slow down with the engine's RPMs. Two flyweights are attached to the spinning shaft. When the shaft spins fast, the flyweights fling out; when the engine slows down, the centrifugal force on the weights becomes less, and the speeder spring pulls the flyweights back in (Figs. 4-8A and 4-8B).

The flyweights are connected to a high-pressure oil valve: the flood gate. In this example, when the engine slows down, the spinning shaft also slows down, which reduces centrifugal force on the flyweights and the weights come together under the pressure of the speeder spring. When the flyweights come together, they move a pilot valve (flood gate), then oil floods into the prop hub and pushes against the hub spring. The prop blades move to a lower pitch and this reduces prop drag. With less prop drag, the engine speeds back up to the setting that the pilot placed on the prop control.

All this happens instantly. Usually the pilot does not know it happened. Study and understand the prop governor system on the airplane that you fly because system configurations might vary from model to model. (Get to know the whole airplane.)

Governing a chandelle

The governor is in constant operation and continuously adjusting to even minor changes in RPM. Take a look at the propeller blade angles and prop governor actions during a common flight maneuver, the chandelle. A chandelle is a maximum-perfor-

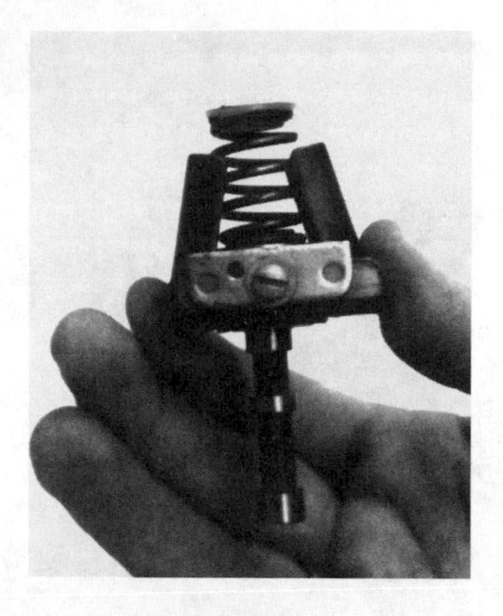

Fig. 4-8A. *The speeder spring is small in comparison with the hub spring. In this photo, the flyweights (held in position by the thumb and forefinger) are closed as they would be in a slow-speed condition.*

Fig. 4-8B. *The same speeder spring and flyweights now placed in the proper location inside the governor. The flyweights are also shown retracted, as if at slow speed.*

mance climbing turn where airspeed is sacrificed for altitude. The maneuver starts by setting the propeller to the proper RPM, usually a climb or takeoff position. This setting will be 2,500 RPM on most airplanes with constant-speed props. The prop governor's job will be to maintain 2,500 RPM throughout the maneuver.

The maneuver starts at design maneuvering speed (V_A) with an easily seen object near the horizon on one wingtip for reference. Some airplanes require a slight descent to achieve V_A. During this descent, the engine will tend to increase speed. When the airplane goes "downhill," gravity assists this motion and not as much energy is required from the engine. With the engine less burdened at the same power setting, the engine RPM increases to perhaps 2,600 RPM.

This rise in RPM is instantly sensed by the prop governor because the spinning shaft inside the governor and the crankshaft are geared together; most systems use the camshaft. The increase in RPM produces an increase in the governor's shaft speed, which increases the centrifugal force on the flyweights. The flyweights will tend to move out (Figs. 4-9A and 4-9B), and this pulls the pilot valve up, releasing oil pressure from the back side of the piston.

Fig. 4-9A. *The governor's flyweights in the extended position, as they would be at a fast RPM.*

Fig. 4-9B. *The spring and flyweights in the governor, this time with the flyweights thrown out, as would be the case at high speeds due to centrifugal force.*

With no oil pressure in opposition, the large hub spring then squeezes out the oil by moving the piston aft, and simultaneously altering the prop blade angle to a higher pitch. The higher pitch produces higher drag on the prop, which labors the engine just enough to slow it back down to 2,500 RPM.

Later in the chandelle, the nose is raised above the horizon, and the airplane struggles against gravity to reach a maximum climb. The engine must work against gravity during the climb to pull the airplane higher. This extra burden causes the RPMs to be reduced, perhaps to 2,400 RPM. The prop governor instantly swings into action.

The governor's shaft slows down so that the flyweights have less centrifugal force and fall back in. This moves the pilot valve into a position where the high-pressure governor-pump oil is allowed to pass to the aft side of the hub's piston. The oil overpowers the hub spring and the piston moves forward. This twists the propeller blade to a lower angle of attack. The lower angle reduces drag, and the RPMs move back to 2,500.

What a chain reaction this is! It sounds like the song, "the ankle bone is connected to the leg bone, the leg bone is connected to the knee bone, the knee bone . . .." The progression is crankshaft, camshaft, prop governor, flyweights, pilot valve, oil pressure, hub-piston movement, blade-angle change, drag change, RPM change. Again, please check your airplane's governor operation. Some do not work exactly as described here, but all work to constantly maximize propeller efficiency.

FEATHERING

In addition to maximizing propeller efficiency, the multiengine system must also have a mechanism that feathers and unfeathers the propeller (Fig. 4-10). This mechanism does not rely entirely upon governor pump pressure, but rather a combination of oil pressure and centrifugal force. It is not a good idea to use oil pressure alone to feather the prop blades because if the governor oil pump failed, the propeller could not feather.

Recall from chapter 3 regarding engine-out procedures that when a prop needs to be feathered, it *really* needs to be feathered; therefore, a truly dependable system must be used. Nothing is more dependable than the basic forces of physics, such as centrifugal force. Propeller blades that feather have counterweights attached to the shank of the blade. The counterweights are often hidden by the spinner.

Fig. 4-10. *This propeller is in the feathered position. It is easy to see that a propeller in this edge-on position will greatly reduce drag.*

During normal operation, the turning prop blade acts as an airfoil. We know that variations of pressure act on the surface of the blade. Taken together, these pressures tend to turn the blade angle to a lower pitch, called *aerodynamic twisting force*. Centrifugal force opposes the aerodynamic twisting force. That opposition tends to throw through the blades to a high pitch.

Usually these two forces cancel each other out, and the combination of oil pressure and a hub spring moves the blades. When the prop blade angle is low, the counterweights are tucked in and have little effect. When the blade angle gets higher, the counterweights rotate with the blade into a new position. This new position places the counterweight's center of gravity farther out from the spinner.

This essentially gives the counterweights an arm, and the centrifugal force is magnified through this arm. The magnified centrifugal force becomes stronger than the aerodynamic twisting force, and the blades move into an even greater pitch angle. The greater the pitch angle, the closer to the feather position.

Many systems also have air charges, or even an additional spring, to help push the blades all the way to the feathered position. These systems will cause the prop blades to feather within only a few seconds of placing the prop control into the full-feather position.

Featherable propellers also have locking pins to prevent feathering below a certain RPM. Consult the operating manual for the proper RPM setting during engine shutdown. Using the pins properly will prevent the blades from feathering during start-up and shutdown.

Autofeathering

Some systems employ an autofeather feature, usually only on high-performance turboprop airplanes. When an autofeather system is in use, the propeller will automatically move to the feathered position when an engine fails, without the pilot moving any controls. When the system is turned on, an electrical sensor compares engine power output to the position of the power levers (throttles). If the power output does not match what should be produced at a given power-lever setting, the system activates, and the prop feathers.

This system cuts down the time between engine failure and feathered prop because it effectively cuts out the time a pilot would take to decide to feather, and then manually move the prop control to feather. The system puts the airplane in the best low-drag configuration as quickly as possible, which frees the pilot to control the airplane. If an engine does fail while in flight, and the prop is feathered, the best thing to do is land as soon as practical.

Restart

Instructors routinely feather props at altitude during a training flight to teach students how to fly with only one engine. The engine has to be restarted, which requires a system that will bring the propeller away from feather and into normal blade angles. When feathered, the blade angle is never precisely 90°. Instead, the blade's farthest travel is approximately 85°–87°. This position lets the airflow help the engine starter during an in-flight restart from feather.

To air-start a feathered engine, place the propeller control into the normal operating range. Crank the engine. As the engine begins to turn, and the blades eventually windmill, the blades will move out of feather within a few seconds. An *accumulator* on certain airplanes helps the governor move the blades out of feather. The mechanism accumulates oil pressure to be deployed at the moment of unfeather. Accumulators (Fig. 4-11) use the pent-up oil pressure to bring the blades from feather to windmill by simply placing the prop control full forward.

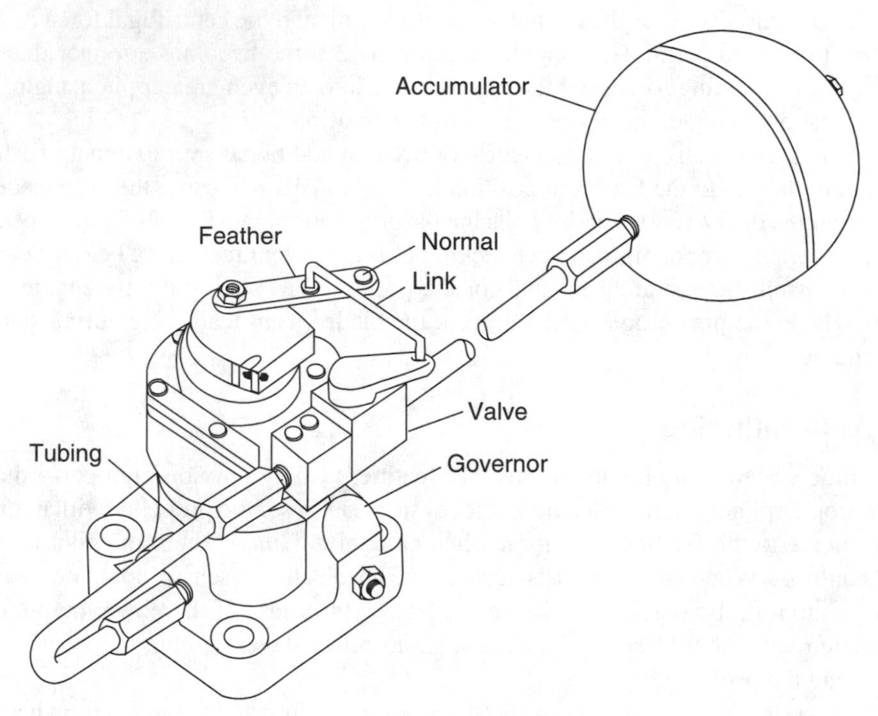

Fig. 4-11. *An accumulator stores up a charge of air and oil. The charge is released to help the governor bring the propeller out of feather.*

REVERSE-THRUST PROPELLERS

Some propellers can throw a thrust force in the "wrong" direction. The blade's direction of rotation does not change, and the angle of attack moves to a negative position. This system is found on turboprops. The reverse-thrust position of the propeller is mainly used to shorten a landing roll or to stop without using brakes on a slick runway. Reverse thrust can be used during taxi to get into or out of a tight spot, or even to back into a parking space. Reverse thrust could also be used in-flight for an emergency descent.

RANGE OF OPERATION

We have discussed three ranges of propeller angles: positive, feathered, and reverse. The feathered position is edge-on and occurs near 90°. The 0° blade-angle position would provide minimal drag and high RPM, but minimal thrust, as well. Normal operations take place from 0°–90°. Takeoff might be 8° and cruise near 12°. The positive-angle range from 0°–90° is the *alpha range*. Reverse thrust is developed when the propeller's blade angle moves to a negative angle. Negative angles from 0° to approximately –20° are in the *beta range* (Fig. 4-12).

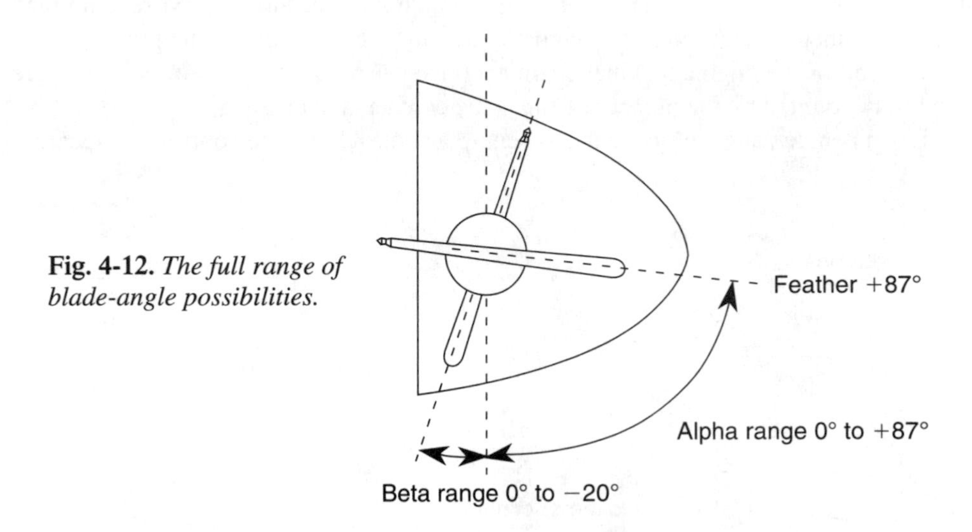

Fig. 4-12. *The full range of blade-angle possibilities.*

CHECKING AND SYNCHRONIZING THE PROPS

Always check the prop governor and feathering mechanism during the pretakeoff checklist. Place both prop controls full forward. Advance the throttles and match the RPMs. This test can be done during the engine run-up, so use the run-up RPM setting. Slowly pull back one prop control until the RPMs drop by 100, then bring back the other prop control until the RPMs drop by 100. This proves that the governors are operating.

Look at the position of the left and right prop control levers. Chances are good that they are not exactly lined up; they are *split*. The cables that connect the prop control lever to the prop governor are never exactly the same for each engine. This means that if you place the two prop controls exactly together, the RPMs from both engines will not be in synchronization.

The props are synchronized in flight by adjusting the controls to the proper split. It is best to know what the proper split is before leaving the ground so that you know what to expect while in flight. By performing this test on the ground, you determine the prop control split for synchronization. Check the airplane's operating handbook for the exact procedure.

Oscillations from unsynchronized propellers will be unnerving. To put the props in sync, set the desired RPM on the left engine first. Set the right-engine RPMs to match the left-engine RPMs. Tachometers are not always perfect, so after matching the RPM numbers, use your ears. You will be able to sync the props by sound with a little practice.

Automatic synchronization

A multiengine airplane might have a device that will synchronize the propellers automatically (Fig. 4-13). The system uses a much more accurate RPM reading than just a tachometer, determined by magnetic pick-ups on the engine. The pick-ups turn at the speed of the engine and make contact once per revolution. The revolutions are accurately counted by the number of times the contact is made.

Upon contact, an electrical pulse is sent to a control box. The control box receives

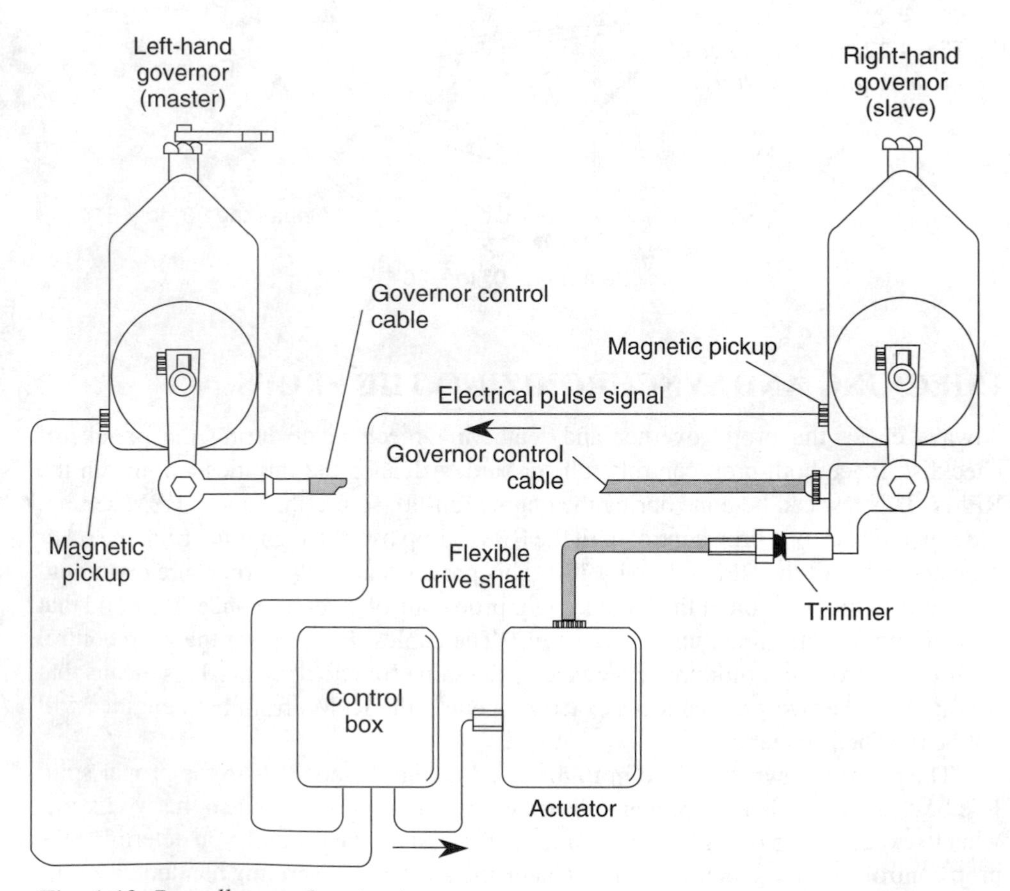

Fig. 4-13. *Propeller synchronization system.*

pulses from both engines and compares them. The left engine is usually considered the master engine; the right engine is the slave. The left engine's RPM is used as the starting point, and the right engine is matched to the left.

A signal is sent to an actuator if the control box determines that the right engine is not turning at the same speed as the left. The actuator changes the RPM of the right engine, using the prop governor, until the two engine's speeds are equivalent. Each time the control box sends a signal, the actuator changes the right engine's speed a predetermined amount—a *step*. It might take several steps to completely synchronize the propellers.

The right engine follows the left engine only through a narrow speed range. If the left engine's speed changes drastically, the right engine will not follow. This is a good idea. What if the pilot feathered the left prop, and the prop sync system feathered the right prop so they would be the same? The pilot can turn the prop sync system off for takeoffs and engine-out operations.

PROPELLER ICE CONTROL SYSTEMS

An airplane with two engines cannot automatically fly safely into any weather. Most light-twin airplanes have absolutely no anti-ice or deice equipment; however, larger airplanes will likely have an ice control system. According to federal regulations, airplanes without ice protection should never be flown into a known-icing condition. Don't let an ice protection system give you a false sense of security. You must become completely familiar with the limitations of an ice protection system.

Ice can form on propeller blades and cause real problems. First, the airfoil shape can become distorted by the ice accumulation, which reduces the blade's ability to produce thrust. Second, the ice might form on the blade unevenly, which unbalances the blades; a destructive vibration can start. Eventually, large ice chunks will be thrown off the blades, which could damage other parts of the airplane.

Two systems have been designed to prevent the ice from accumulating on a propeller: fluid and electrical. The fluid system employees anti-icing liquid, commonly isopropyl alcohol, that is pumped from a tank to the propeller hub (Fig. 4-14 on page 82). At the hub, the fluid is passed from the stationary line to the spinning propeller through a *slinger ring*. Centrifugal force pulls the fluid from the slinger ring onto the blades.

In some systems, the fluid will travel inside the blade through a narrow strip of rubber, called a *feed shoe*, and be dispersed on the outside of the blade.

The electrical system uses a *hot blade* to keep ice clear. These propeller blades have a heating element that is either inside the blade or mounted externally. The airplane's electrical system power is transferred from stationary wires to the spinning prop by using contact brushes that ride on a rotating ring. When the electrical current makes the leap to the prop, the heating elements are activated just like the heated wires that melt frost off a car's rear window. The prop ice must melt uniformly to prevent vibration; therefore, the intensity of the heat is controlled.

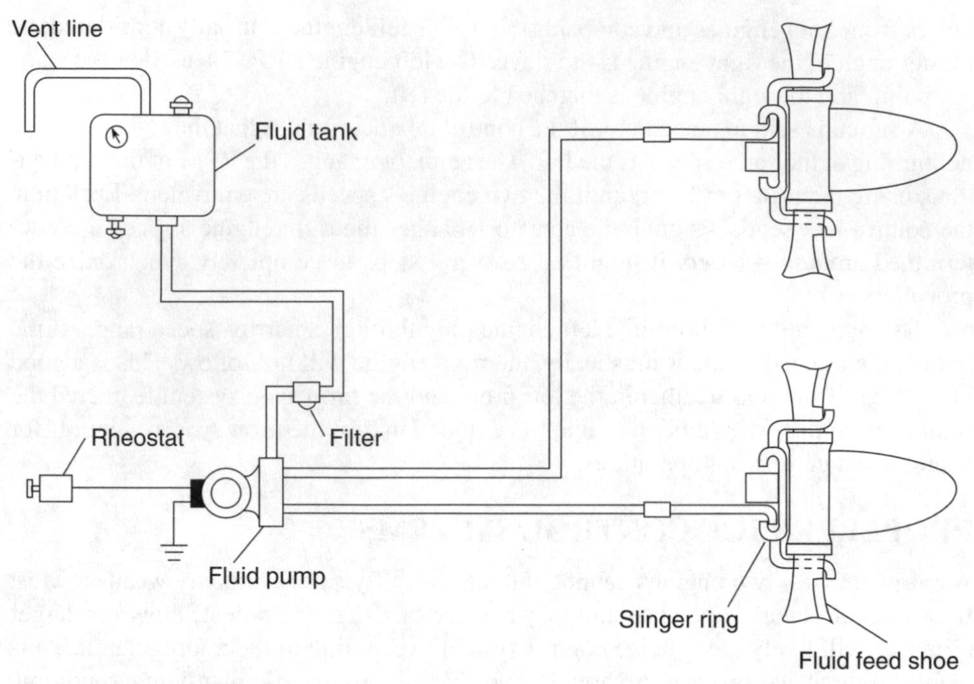

Fig. 4-14. *A propeller's liquid deice system.*

5

Multiengine fuel systems

BASIC COMPONENTS OF A SINGLE-ENGINE AIRPLANE FUEL SYSTEM ARE THE same on a multiengine airplane, except that there might be more fuel tanks and the ability to crossfeed fuel between tanks. Many multiengine airplanes have more than one fuel tank in each wing because fuel consumption is higher: two engines, each with more horsepower, vs. one engine with lower horsepower. Additional fuel will also extend the range of the airplane. Extra tanks might also be a space requirement or meet a structural need.

The tank with the greatest capacity is usually the main tank, and all others are auxiliary tanks. Auxiliary tanks can be in the wing, wing tips, or fuselage. The main and auxiliary tanks are filled separately, are vented separately, and might feed the engine separately.

Some systems require fuel to be transferred from a wingtip tank to an inboard main tank. The quantity of fuel in the main tank must be low enough for the fuel transfer to occur. Also, adequate time must be allowed for the fuel transfer to take place. Obviously, the pilot must fully understand the system to accomplish proper fuel management.

Fuel normally stays on one side of the airplane; the fuel from all left-wing tanks is used in the left engine, and all right-wing fuel is used in the right engine. This works fine until one engine fails in flight. If the failure occurs while flying in heavy IFR weather, you might need all the range you can get. You might determine that the fuel supply for the operating engine will run out before you can land. In this situation, the fuel from the dead engine side can be transferred across the airplane to the good engine.

CROSSFEEDING

Study the airplane's operating handbook to determine the exact valve settings to crossfeed and how to maintain the crossfeed system. Because the crossfeed line will not be used as often as the normal lines, the crossfeed line is very susceptible to water and debris accumulation. Many airplanes will have a separate place to draw a crossfeed-fuel sample. Find the crossfeed drain sump and take a fuel sample before each flight.

Test the crossfeed operations on the ground. While taxiing out for takeoff, switch the valves so that crossfeed fuel goes to the engines, and allow enough time for the fuel to arrive at its destination engine; however, before takeoff, switch the valves back to the normal position and allow time for that flow to take place. This ensures that the crossfeed will work when airborne, and a single operating engine will continue to run in the critical situation.

The valve system on light twins is usually very simple. The settings are ON, OFF, and CROSSFEED (Figs. 5-1 and 5-2). Normally, both valves are on, meaning that fuel is moving from the left tank(s) to the left engine, and fuel is moving from the right tank(s) to the right engine. When an engine fails and the need for crossfeed arises, you must know exactly how to position the valves.

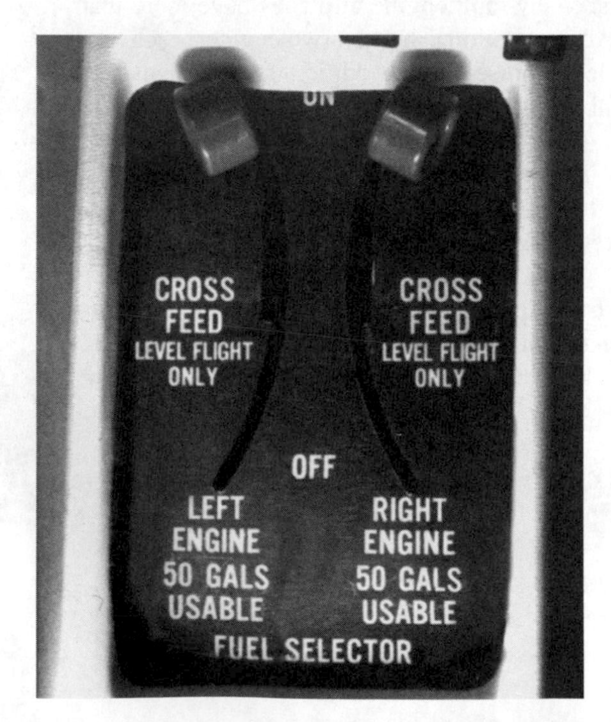

Fig. 5-1. *Fuel selector valves for a multiengine airplane with one fuel tank in each wing.*

Fig. 5-2. *Fuel selector panel for a multiengine airplane with main and auxiliary fuel tanks.*

If the left engine fails and cannot be restarted while the fuel selector valve is in the ON position, it will be important to turn the fuel valve to OFF. The airplane's operating handbook will have a securing-engine checklist, which is used as damage control. You do not want to send fuel to the problem-engine. Follow the manufacturer's recommendations to turn off fuel pumps, and stop the flow of fuel. If the flow is not stopped, a fire could start, turning the engine-out emergency into a catastrophe.

After the fuel flow is stopped to the dead engine, fuel consumption to the good engine will increase because the higher power setting is required to maintain level flight. The pilot should fly to the nearest suitable airport and land. It might become necessary to use fuel on the dead-engine in the good engine. Turn the good engine's fuel selector from ON to CROSSFEED, and turn the dead engine's selector to OFF (or as prescribed in the operating handbook). You cannot normally mix fuel from both left and right; fuel comes exclusively from one side or the other.

Crossfeeding fuel will also help balance the airplane. If one side's tank burns down to near empty, while the other side is near full, aircraft control will become awkward. This condition could even be dangerous if a crosswind landing must be made.

FUEL PUMPS

Multiengine pilots must also become familiar with fuel pump operations. Many of us learned to fly in high-wing airplanes that did not have an electric fuel pump to worry about. Gravity did the work and we never thought much about it. Pilots of low-wing airplanes are accustomed to fuel-pump switches. Most light-twin airplanes are low-wing; therefore, understanding the pumps are part of understanding the airplane.

Fuel pumps are either engine-driven or electric. Engine-driven pumps are geared to the engine and turn anytime the engine is turning. These pumps move the fuel under pressure through the line and to the engine. The pumps very rarely fail, but if they fail, the engine could stop due to fuel starvation. Electric fuel pumps are provided as a backup system.

The electric pump operates only when it is switched on; power comes from the airplane's electrical system. The electric pump is first used for engine start-up because the engine-driven pump cannot supply fuel pressure. This is the only time when the electric pump is working and the engine-driven pump is not working. All other times, when the engine is running, the electric pump is used as a supplement to the engine-driven pump.

Pumps should be tested prior to takeoff. If the electric pump is used to start the engine, turn off the pump after the engine begins to run. Whenever you turn off a fuel pump, immediately look at the fuel-flow or fuel-pressure gauge (Fig. 5-3). Depending on the airplane, a noticeable change in pressure might be seen. If the fuel pressure falls below the green arc of the gauge, and stays there, an engine-driven fuel pump has failed. You must look at the fuel-pressure gauge; if you do not catch the pressure problem early, the engine will stop.

Theoretically, when both electric and engine-driven pumps are functioning, there will be a higher pressure than when the engine-driven pump alone is working. But two-

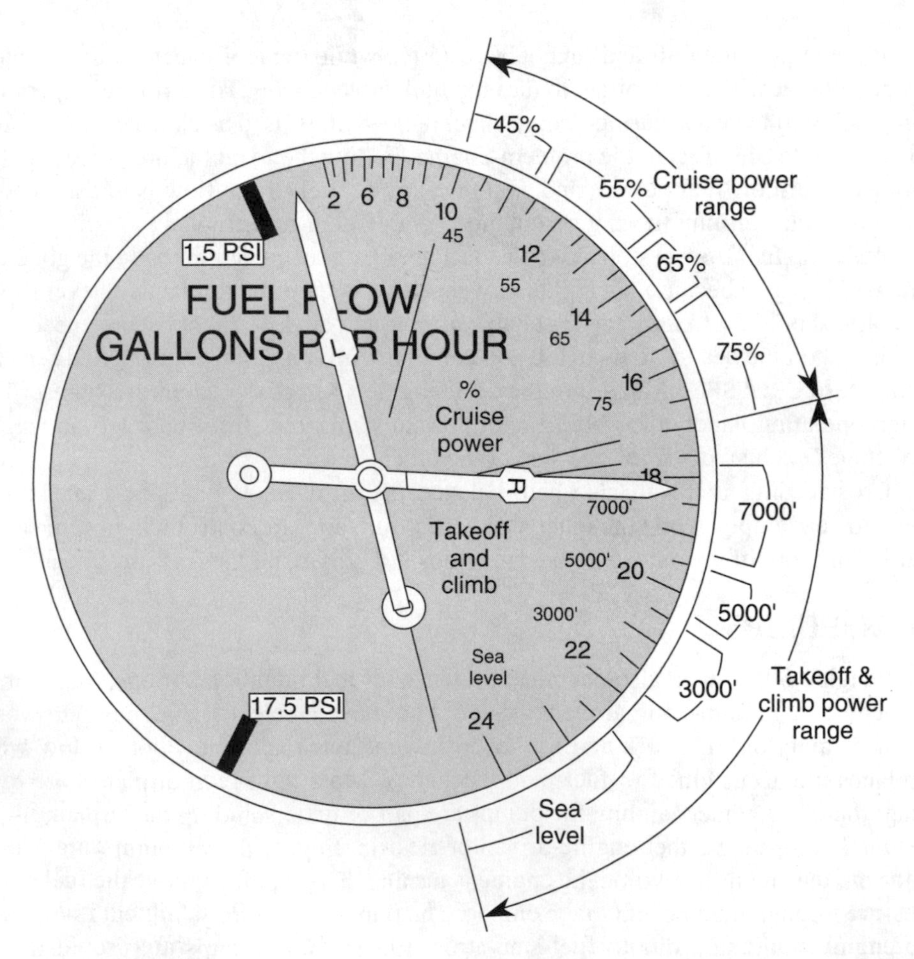

Fig. 5-3. *Typical fuel flow meter.*

pump and one-pump pressure should be within the green arc of the fuel-pressure gauge. When the electric pump is switched on, a rise in pressure should be seen, and this will verify the operation of the electric pump.

There are crucial times when an engine failure would be particularly dangerous, for instance, takeoff. All fuel pumps should be operating during crucial phases of flight; if one pump fails, the other continues to push the fuel to the engine. After a crucial phase of flight has passed (arriving at cruise altitude), the electric pumps may be turned off.

Never turn both pumps off at the same time. Turn one engine's pump off, and watch the fuel-pressure gauge to ensure that the engine-driven pump is still maintaining pressure in the green arc. When that engine's fuel flow is stable, turn off the second engine's pump, and watch the pressure on that side.

Improve your fuel-management safety margin by locating the nearest airport before turning off any pumps or switching any tanks. Switching tanks over an airport is an ideal procedure en route. One decision is already made if you must seek safe haven due to a fuel system malfunction.

FUEL WEIGHTS

Fuel management usually addresses weight and balance and range concerns, but in many multiengine airplanes it is also a structural topic. A certain amount of fuel might be required in the airplane for the airplane to be strong enough for flight. The fuel in the tanks can actually make a wing stronger. Which soft drink can is easier to crush? The full, unopened can, or the empty can? The empty can, of course. Which fuel tank is easier to bend? The empty one. Nobody would deliberately allow a fuel tank to run dry, but there might be danger in even letting it run low. There are two important terms to know about: *minimum fuel for flight* and *zero fuel weight*.

The minimum fuel for flight is a fuel reserve, but not for the normal reasons. Most pilots can tell you about the VFR day, VFR night, and IFR fuel reserve regulations, but the writer of these regulations did not have the structure of the airplane in mind. These regulations are written to allow lost pilots to find an airport before the fuel runs out. Minimum fuel for flight is different.

Some airplanes must have a particular weight of fuel, or greater, to remain structurally sound. The fuel usually rides in the wings. The design of the wings allows for the weight of fuel. When the fuel runs low, the wing becomes too light, and the wing will rise higher. The lack of fuel allows the wing to flex too far for safety.

The junction where the wing meets the fuselage is like your shoulder joint when your arm is extended. When the wing has the proper weight, it's as if your arm were held straight out from the shoulder, parallel to the ground. When the wing is light, lift pulls the wing up, like raising your outstretched arm. The joint only has so much play, and structural damage (wing bending), or failure (wing breaking off), might occur if the joint is pushed passed the limit of play. When flying an airplane that lists a specific minimum fuel for flight, running out of gas is the least of your problems. It is more dangerous to run low than to run out.

Zero fuel weight is also related to structure. If an airplane has a zero fuel weight, or zero fuel condition, the amount of payload other than fuel is limited. Zero fuel weight is the maximum an airplane can weigh without considering the weight of usable fuel. Figure 5-4 is a sample weight and balance form from a typical light-twin airplane.

The zero fuel condition of this airplane is 3,500 pounds. This means that the total weight of the airplane, pilot, passengers, and luggage cannot exceed 3,500 pounds. The airplane's maximum takeoff weight is 3,900 pounds. In other words, it is possible to overload the cabin and fuselage without exceeding the maximum gross takeoff weight. Multiengine pilots must pay close attention to this fact so they do not fly too long and go below the zero fuel weight.

WEIGHT AND BALANCE LOADING FORM

Model *Duchess 76* Date ______________

Serial No. ______________ Reg. No. ______________

Item	Weight	Mom/100
1. Basic empty condition		
2. Front seat occupants		
3. 3rd & 4th seat occupants or bench seat occupants		
4.		
5. Aft baggage		
6. **Sub total** zero fuel condition (3500 lbs max.)		
7. Fuel loading (gal.)		
8. **Sub total** ramp condition		
9. *Less fuel for start, taxi, and takeoff		
10. **Sub total** take-off condition		
11. Less fuel to destination		
12. **Landing condition**		

*Fuel for start, taxi, and takeoff is normally 16 lbs at an average mom/100 of 19.

Fig. 5-4. *Sample weight and balance loading form. Item number six on the form refers to a "zero fuel condition" of 3,500 pounds maximum.*

Figures 5-5, 5-6, and 5-7 illustrate the complexities of multiengine fuel systems. If you are more accustomed to airplanes without fuel pumps and only an on/off fuel valve, familiarize yourself with all aspects of a complex airplane's fuel system, and develop routine operating habits that foster safety.

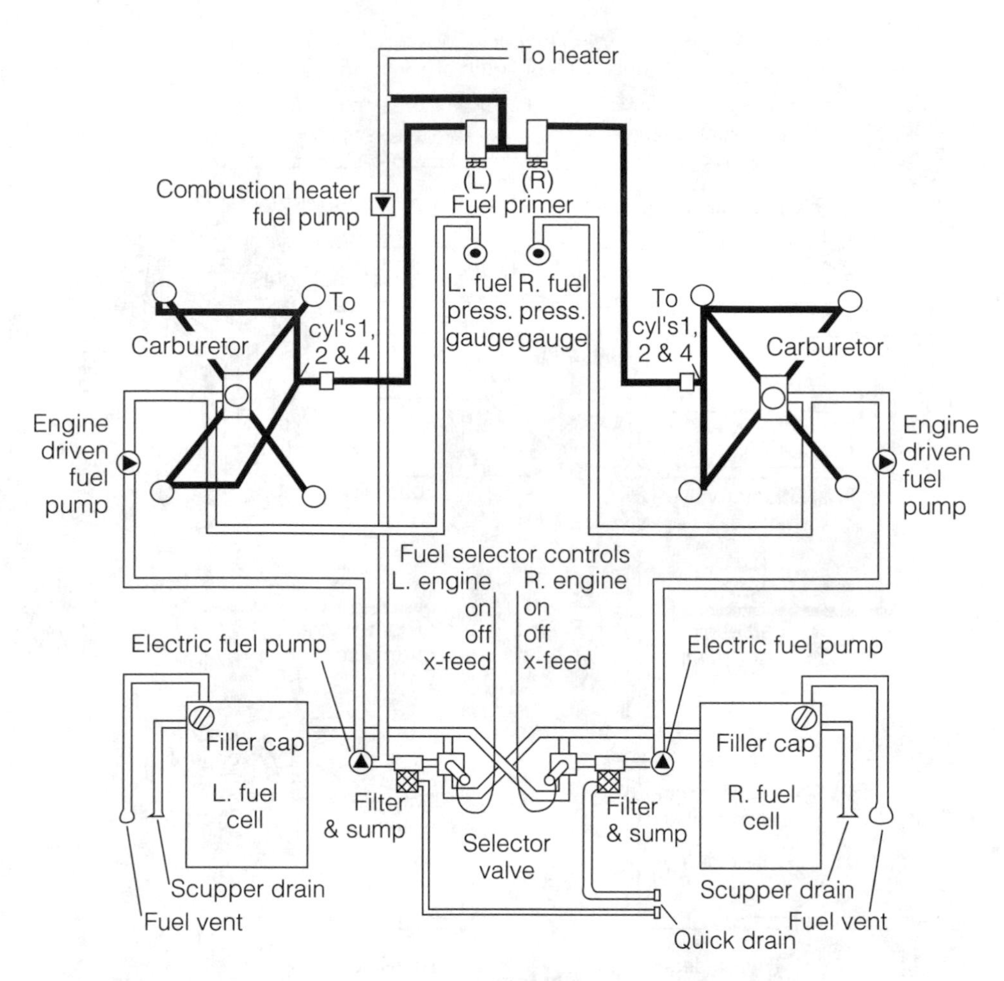

Fig. 5-5. *Piper Seminole fuel system. Note the crossfeed, the cylinders that are primed, and where fuel for the cabin heater comes from.* Piper Aircraft Corporation

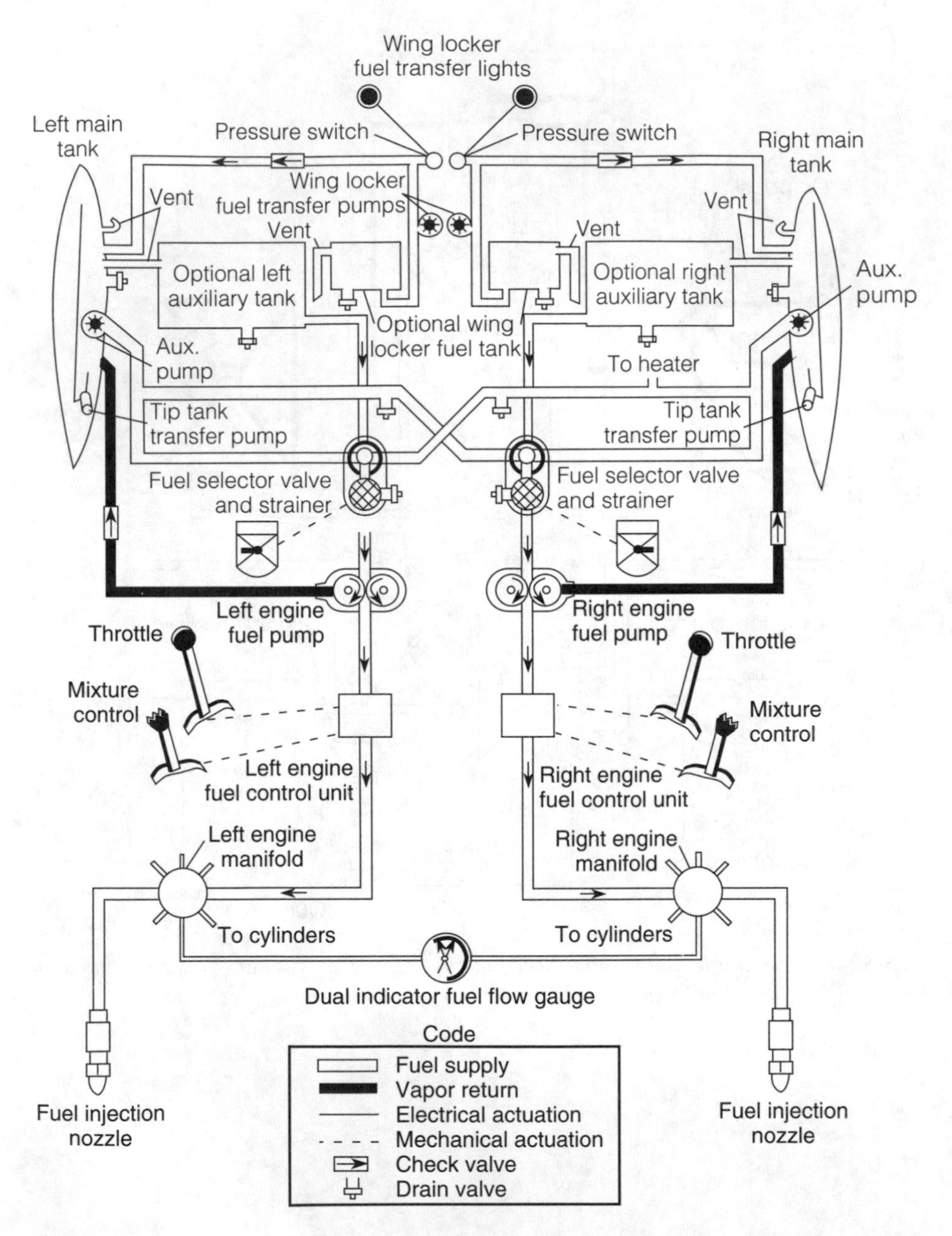

Fig. 5-6. *Cessna 310 fuel system with main, auxiliary, and "locker" tanks, and the wingtip fuel transfer pumps.* Cessna Aircraft Company

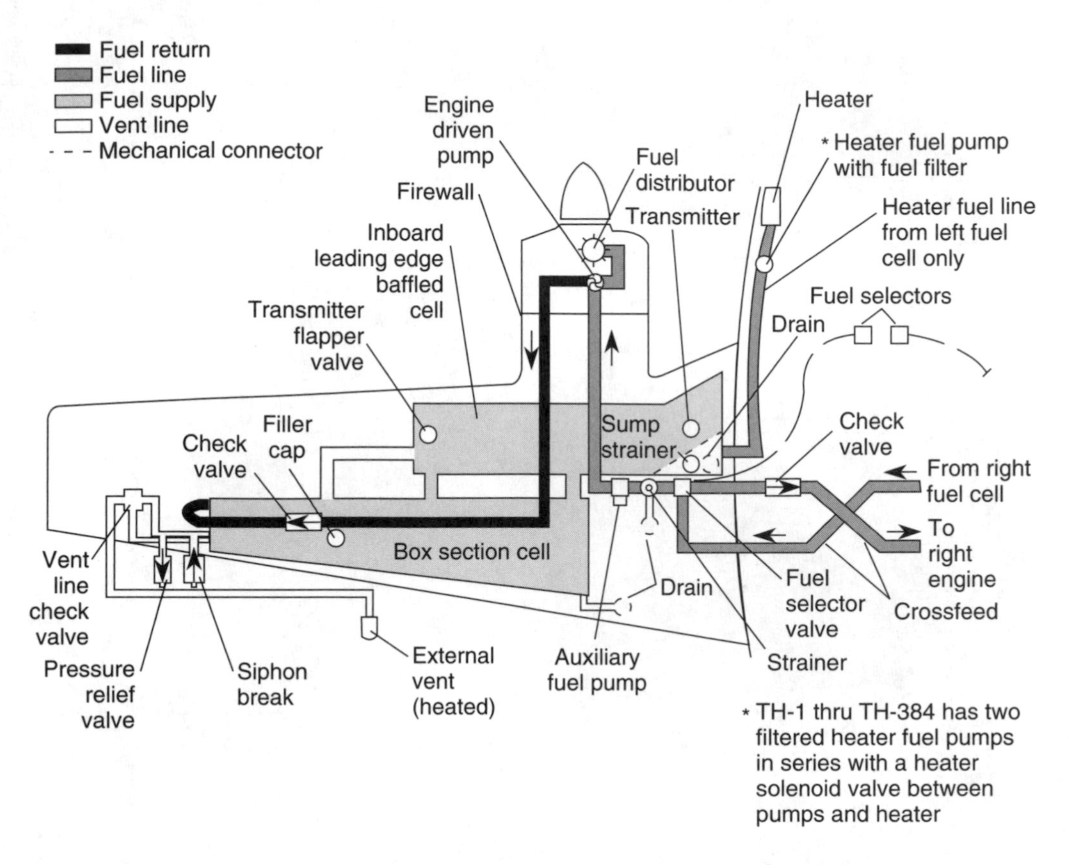

Fig. 5-7. *Beechcraft Duchess fuel system. Note the position of the engine-driven and electric auxiliary fuel pumps and the crossfeed.* Beech Aircraft Company

6
Multiengine electrical systems

MANY PILOTS HAVE A LIMITED UNDERSTANDING OF THE AIRPLANE'S electrical system because they have a poor foundation of knowledge about electricity. Pilots back off after taking one look at all those circuit breakers, wires, and electrical schematic diagrams. All pilots need a working knowledge of the airplane's electrical system in order to diagnose in-flight electrical problems and troubleshoot the system.

CURRENT KNOWLEDGE

The first step in understanding electricity is to have some concept about what it actually is. The good news is that electricity is not mysterious, and it is really easy to grasp. Recall from physical science classes what you learned about atoms and the parts of the atom. My favorite part was always the electron because it flew around the stationary center part called the nucleus.

An atom becomes more complex when more atoms start orbiting the nucleus. The electrons orbit in layers. Each layer is farther from the center. The electrons that travel in that farthest layer can escape the grasp of that atom and travel freely to other atoms.

When free electrons move in a pattern from one atom to the next, an *electric current* is produced. Some atoms will give away electrons easier than others. Atoms from elements that give up electrons freely are called *conductors*. Atoms that tend to hold their electrons are called *insulators*. An electric wire will be made of material that conducts, or passes, the electron flow. The covering that is wrapped around the wire is an insulator that keeps the flow on the proper course.

How does electricity move so fast? When you hit a light switch, the light instantly comes on. How did the electron move from the switch to the light so quickly? The fact is that a single electron did not move that fast. In reality, the electron at the switch started a chain reaction of electron flow down the wire to the light.

Imagine a length of pipe with ping-pong balls stacked up one after another inside the length of the pipe. If you were to push a new ping-pong ball into one end of the pipe, a ping-pong ball at the far end of the pipe would be pushed out. The inserted ball does not have to roll all the way down the pipe in order for a ball to fall out. The inserted ball pushed all the balls that were already in the pipe; therefore, the end ping-pong ball fell out of the pipe at the exact instant that the new ping-pong ball was pushed in.

In this analogy, the ping-pong balls are free flowing electrons. The pipe is a wire's insulation. When an electric switch is turned on, one electron pushes electrons that were already in the wire until the electron at the far end of the wire is pushed into the light bulb. All electrical devices work the same way, whether the device is an electric razor, or an airplane landing light, or a transponder.

Forcing the issue

For the current flow to take place, a force must move the first electron in the chain. The ping-pong ball was pushed into the pipe. The push is called *electromotive force*, which can be thought of as electrical pressure. This pressure is measured in *volts*. If the ping-pong ball is gently placed into the pipe, a low voltage would exist; if the ball were thrown into the pipe, a high voltage would result.

The number of electrons (ping-pong balls) that pass by a certain position along the wire in a given amount of time is the *electron flow rate*. This rate is measured in *amperes*. When a large number of electrons flow, there will be a greater number of amps.

The electrons will also encounter some resistance as they move. *Electric friction* will slow the electron's flow. This resistance is measured by a unit called an *ohm*.

Grounded

The flow of electricity must move in a loop. If electrons flow into a light bulb, there must be a path for the electrons to return from the light to the power source. In airplanes, wire (and weight) is often saved by using the frame of the airplane as the return-wire from most electrical equipment. This means the electrons flow from a power source to an electrical device through a wire. When leaving the device, the electrons are led to a *ground* wire that is attached to the airplane. The electrons flow through the airplane, back to the power source, and the loop is completed.

The battery or an *auxiliary power unit* (APU) provides power before an engine starts. When an engine has started, a generator or alternator will be turned by a gear off the engine. As the generator or alternator "comes on-line," the electrical system takes over from the battery. The battery is out of the loop, and its charge is no longer utilized.

Flowing

A generator need only be turned to generate an electric current flow. This seems simple enough, but the generator's current flow is affected by engine RPM. This can be annoying while taxiing at low RPM or during final approach with the engine power pulled back. Delicate electric circuits can be damaged by wavering tides of electrical current coming from a generator.

An alternator, on the other hand, delivers a predictable and equal flow, regardless of RPM. The alternator uses a *magnetic field* to produce the required current. (The magnetic field is triggered by a small electric current.) Alternators are preferred more than a generator for the equal current flow.

Each engine of a multiengine airplane has an independent electrical system. In most cases, one engine's alternator electrical system is capable of supplying sufficient electric power for the entire airplane's. It is a great feeling to know that if one alternator fails when flying IFR, the other alternator should take up the slack and you fly home safely.

SYSTEM SAVVY

When both engines and alternators are running properly, the electricity drawn from the alternators is shared equally from each side. Figure 6-1 illustrates a simple generic multiengine electrical system. Even though there are two engines, two starter motors, and two alternators, there is only one battery.

The battery is grounded to the frame of the airplane to accept returning electron flow from devices located all over the airplane. The electron flow from the battery will have a series of switches, contacts, or solenoids to pass through to reach the starter motors and the bus bar. When the engines are started, electrons are allowed to flow from the battery by closing the battery switch, known as the *master switch.*

The current reaches the starter motor when the starter switch for that particular engine is closed. The starter switch can be activated by the pilot using a toggle, a button, a key, or a rocker switch. The current enters the starter motor, then passes out to the ground, and goes back to the battery. When a loop of current is connected, the engine turns over. The engine that is closer to the battery should be started first. When the electrons have a shorter distance to travel, there will be less resistance and less drain on the battery.

When the engines have started running, the alternators will be turning because they are geared to the engine. Recall that solely turning the alternator will not produce a current flow. A small amount of current must first enter the alternator to activate the magnetic field circuit.

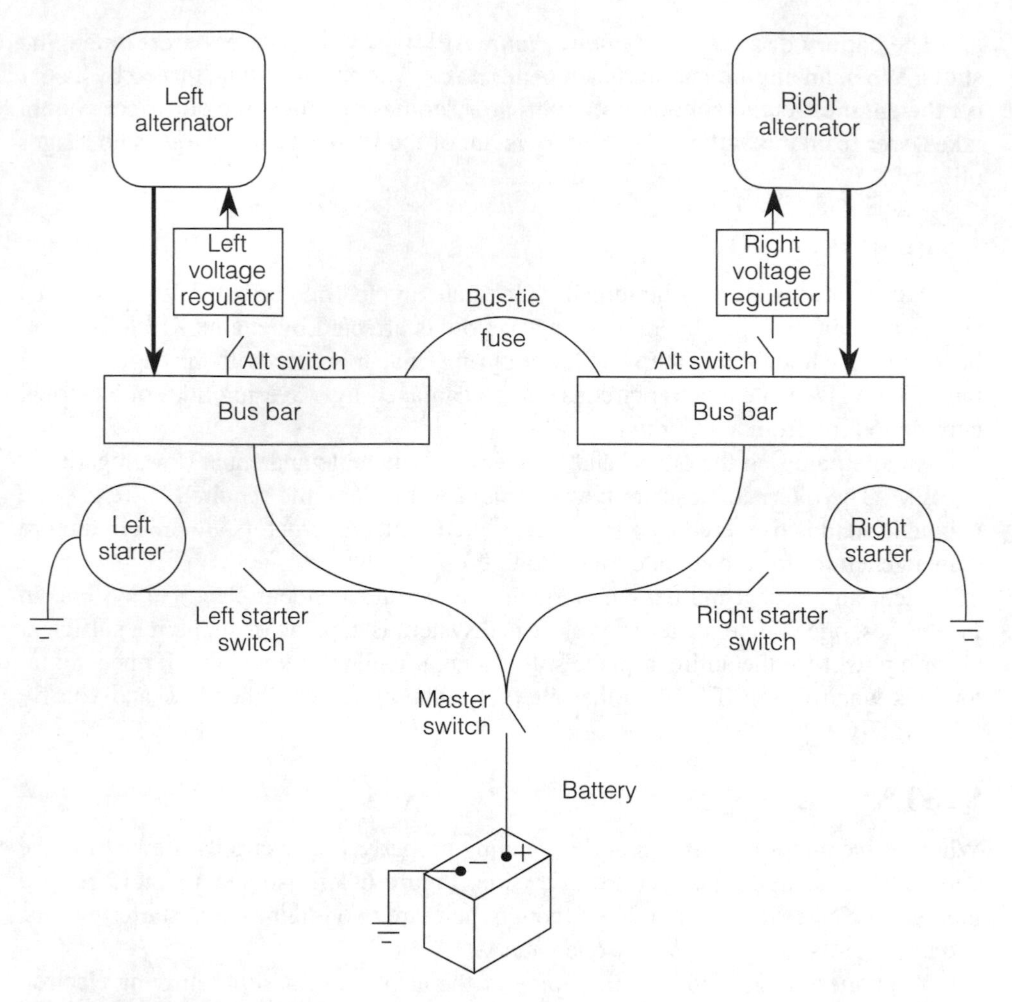

Fig. 6-1. *Generic multiengine electrical system diagram.*

The bus to alternator lines in Fig. 6-1 indicate a small wire with a small current flow. This small current is taken off the bus bar and fed into the alternator. The small current excites the field circuit, which couples with alternator rotation to produce a larger current. The alternator to bus lines represent larger wires that carry the larger current back to the bus bar.

BUS STOP

The *bus bar* is the junction for all other electrical devices in the airplane. A main power line will branch off to supply electricity to individual homes in a neighborhood. The bus bar is a power line, of sorts, that branches off to the lights, flaps, radios, turn coordinator, and all other airplane electrical devices.

The bar is lined with circuit breakers. Each breaker is like a traffic cop. The cop lets in only the proper amount of current to the device. If the current flows too fast or too strong, the cop will stop the electronic traffic flow to the device, which protects the device. The circuit is broken when the breaker opens and interrupts the flow of electrons.

Certain multiengine airplanes utilize a common bus bar. Figure 6-1 illustrates a system with two bus bars. One alternator feeds one bus bar. If an alternator fails, the remaining alternator can provide current to both bus bars through a *bus-tie fuse*.

CONTROL

A certain amount of current from the alternator is recycled to continually supply the alternator's field circuit current. This can lead to a problem. If the field circuit current gets too large, the output of the entire alternator could become too large. This would send more power around the loop back to the field circuit, which would increase the alternator current to even higher levels; hence, the system starts to "feed itself."

Alternator current levels might get so high that circuit breakers will start popping. The electrical devices will shut down, and the alternator will run away. A *voltage regulator* is placed in the system to prevent this from occurring. The regulator acts like a funnel in the field circuit line. It only allows a certain amount of current through. Even if a larger current enters the funnel, only the proper amount exits to the field circuit. The entire alternator's output is under control when the field circuit is under control.

The voltage regulator (funnel) can also widen its opening when the system needs more current flow. Demand is the greatest at night when all the interior and exterior lights are turned on in addition to radios and all other electronic devices. The voltage regulator is able to sense the increased demand and increases the flow to the field circuit, which causes an increase in total alternator output and the electrical demand then is met.

INSTRUMENTATION

The pilot can observe the electrical system through the electrical monitoring instruments. Some airplanes have load meters that display the amps or percentage of capacity. Other systems use ammeters to indicate whether the alternators are supplying sufficient power, or if the battery is tapped for power.

All systems will have a system-failure light or gauge. If the current to the alternator field circuit gets too low or too high at any time, the voltage regulator will cut off the current to protect the system. This dumps the entire electrical load on the battery or the other engine's alternator.

You need to consult the operating handbook to determine the proper course of action to bring the alternator back on-line. If one alternator fails, you need to ensure that the other alternator is indeed providing the proper amount of current, and that the current is being delivered to the proper locations.

Multiengine airplanes provide redundancy, which provides some peace of mind. Two independent alternator systems failing on the same flight would be a very rare occurrence. The multiengine pilot must recognize any evolving electrical problems early, then quickly take constructive action to solve or diminish the problem based upon knowledge about that airplane's electrical system.

7
Multiengine flight training

MANY FLIGHT SCHOOLS OFFER FLIGHT AND GROUND INSTRUCTION THAT leads to the multiengine airplane rating. Check out each school as if you were shopping for any other pilot certificate or rating. Do not shop on the basis of cost alone. Look at the airplanes that will be used for the training. Check out the experience level of the multiengine instructor. Ask questions about the ground instruction and the support facilities.

Ask about insurance requirements for renting multiengine airplanes. Contact an aviation insurance company for details about proper insurance to fully protect yourself. The instruction necessary to get a multiengine rating is usually less than 10 hours, but a minimum 25 hours is common for insurability. This means that you might pass the checkride, but still not be able to rent the airplane for a solo flight.

A multiengine course is a course in disaster management. The initial flight is the only flight where the trainee pilot can get accustomed to the airplane. After the first flight, the lessons are filled with presumed malfunctions, equipment failures, and other assorted in-flight disasters that must be dealt with. Multiengine training is all work and no play. (Adding on a seaplane rating has always been considered more fun. Even though a seaplane rating is serious business, training can be intermingled with fishing trips.)

Any good multiengine course must spend a lot of time on the ground. There is no multiengine rating written test. Ground school might be your only opportunity (besides this book) to understand multiengine concepts. The ratio of flight-to-ground instruction should be at least 1-to-1. If the instructor wants to jump right into the airplane with little or no preliminary ground school about the airplane's systems, aerodynamics, and procedures, look for another school.

The transition to multiengine work will be much easier if you have had some previous experience in a complex single-engine aircraft. If you have never worried about landing gear and prop control settings before, and you try to grasp those while also dealing with two engines, you will be way behind the airplane and the learning curve. Log perhaps 5–10 hours in a complex single-engine airplane first.

What is complex time? To be eligible for a commercial pilot certificate, an applicant must have logged 10 hours of complex airplane time, according to FAR 61.129. The term complex is unusual because it is not defined in FAR part 1. The only reference to the word complex comes from the FAA's practical test standards for the commercial and flight instructor certificates. The publications describe an airplane that must be used for the flight test.

The airplane must have a constant-speed propeller, retractable landing gear, and retractable flaps. This means that a Learjet is not a complex airplane because it does not have a propeller. Engine horsepower is not mentioned.

Complex should not confused with the term *high-performance*. A high-performance airplane, defined in FAR 61.31, has 200-horsepower, or the triple combination of constant-speed prop, retractable gear, and retractable flaps. I train commercial pilot and flight instructor candidates in an airplane that has a constant-speed prop, retractable gear, and retractable flaps, but it only has 180 horsepower; therefore, I am teaching in a high-performance airplane because the regulation has that all important "or" in the sentence. The airplane is also complex.

A flight instructor must make an endorsement in your logbook stating that you have been adequately trained to fly a high-performance airplane before you can log pilot-in-command time in such an airplane. Many students have an endorsement that says "complex" rather than "high-performance" in their logbook. Check your own book. If it says complex, go back and have your instructor change it to high-performance or you will legally be unable to log or act as PIC in that airplane.

Multiengine airplanes used for training are both complex and high performance. For this reason, some students have combined a multiengine rating with a commercial pilot certificate checkride. They receive training and become proficient on commercial maneuvers in a single-engine noncomplex airplane before moving to the multiengine airplane to log at least 10 hours of complex-airplane flight time.

They have two applications for the checkride: commercial certificate and multiengine rating. This is fine except for two possible problems. First, this combination might require the student to move faster than he is capable of. Moving from a Skipper to a Duchess is a big step. The student might be able to fly the airplane after some practice, but his systems experience is very limited. Second, the student will have little or no single-engine complex time, which will be a problem if she goes for a flight instructor certificate.

People pursuing a career as a professional pilot should consider starting multiengine training after obtaining the commercial pilot certificate and instrument rating. This makes things less complicated, promoting building-block training from one step to the next. Seemingly countless variations of training are certainly possible; the emphasis is don't train beyond your capabilities and sacrifice safety for time.

The multiengine rating comes in two types: VFR or IFR. At one time, a person with an instrument rating who later earned the multiengine rating could then fly IFR in the multiengine airplane. A series of accidents led the FAA to change the law. The FAA determined that IFR in a multiengine airplane was more demanding on the pilot than IFR in a single; yet no additional instrument/multi training was required.

I received this letter from the FAA in 1984 regarding the matter (edited):

"On July 2, 1981, a Beech 65-A80-8800 crashed near Madisonville, Texas. The NTSB investigation indicated that the accident might have been caused by excessive air loads generated by nose-up control input by the pilot at high speed, which resulted in in-flight breakup of the aircraft.

"A review of the pilot's records indicated that he had acquired an instrument rating in a single-engine aircraft; however, he had limited experience in the operation of multiengine aircraft in instrument meteorological conditions and had not received instrument training in a multiengine aircraft.

"Consequently, the NTSB recommended that the FAA require all holders of an instrument rating and a multiengine class rating to demonstrate their ability to operate a multiengine aircraft under normal and emergency conditions by reference to flight instruments only as a prerequisite to exercising the privileges of an instrument rating in multiengine aircraft.

"The FAA agrees in principle with the NTSB recommendation as it applies to airplanes; therefore, appropriate flight-test guidance material will be amended to reflect that all new applicants for a multiengine airplane class rating who hold an instrument rating for airplanes will be required to demonstrate pilot competency to operate a multiengine airplane solely by reference to instruments.

"Persons who currently hold an instrument rating for airplanes plus the multiengine airplane class rating will not be affected; however, if the applicant elects not to demonstrate competency in instrument flight, the applicant's multiengine airplane privileges will be limited to visual flight only.

"Multiengine ratings restricted to VFR will . . . (state) . . . AIRPLANE MULTI-ENGINE VFR ONLY. To remove this restriction, the pilot will be required to demonstrate competency to operate a multiengine airplane . . . solely by reference to instruments."

The NTSB investigation of multiengine and related instrument flying accidents led to a recommendation to the FAA that multiengine testing be changed. The FAA took this advice and added three items to the multiengine flight test for IFR privileges:

- Instrument approach with both engines operating.
- Instrument approach with only one engine operating.

- Engine failures during straight and level flight and in turns while flying solely by flight instruments.

This became known as the "first handshake" rule. Multiengine applicants have to declare at the time they first shake the hand of the FAA designated pilot examiner that they are arriving to take either a VFR or IFR multiengine test. The method came about because applicants would say they wanted an IFR multiengine rating. If they failed an IFR item, they would say, "Forget the IFR part, I really only wanted a VFR rating." The rule means that the multiengine flight test is all or nothing.

Decide whether to train for IFR or VFR and tell your instructor. This reemphasizes the importance of obtaining the instrument rating before the multiengine rating.

When investigating flight schools and multiengine instructors, ask to see the flight syllabus to be used. Some schools might not have a published syllabus, but the instructor should have a clear plan of instruction that will take you from proficient single-engine pilot to competent multiengine pilot. The regulations do not require a minimum amount of multiengine dual instruction. For this reason, most multiengine training programs are short, ranging from 6–12 hours of instruction. With a minimum amount of time in the airplane, it is very important that every minute count. If the instructor does not have an efficient plan of action, then look elsewhere.

The following is a sample multiengine course syllabus and a discussion of each item taught. The syllabus that you use might not be exactly like this one, but the topics and maneuvers should be the same. I have used this IFR multiengine rating syllabus a hundred times with great success.

The lessons are not automatically one hour, or even one flight. The student only advances to the next lesson when the completion standards of the previous lesson are achieved. Flight time required depends on the student's ability to grasp the topics, previous experience in complex airplanes, and IFR proficiency.

LESSON 1

Objective: The student will become familiar with the multiengine airplane's systems, controls, and cockpit layout. The student will be introduced to flight at critically slow airspeeds, become familiar with the flight characteristics of slow and stalled flight, and become familiar with the airplane's normal flight characteristics.

Introduce:
Preflight and ground maneuvers
Aircraft systems and airworthiness inspection
Cockpit resource management
Safe engine starting procedures
Taxiing: normal, crosswind, and with differential power
Pretakeoff checklist and systems check
Flight operations
Traffic pattern operations

Four fundamentals:
- Slow flight
- Approach to landing stall
- Takeoff stall
- Steep turns

Heater operation
Autopilot operation
Manual gear extension
Simulated engine failure en route (instructor demo)
Drag demonstration

Completion standards: The student will be able to perform all the listed ground procedures with instructor assistance. During takeoff and landing, the student will demonstrate good directional control and maintain liftoff, climb, approach, and touchdown airspeed within 10 knots of the correct airspeed. Straight-and-level flight, climbs, and descents will be performed while maintaining assigned airspeeds within 10 knots, rollouts from turns within 10° of assigned headings, and specified altitudes within 100 feet. The student will be able to demonstrate the correct flight procedures for maneuvering during slow flight, steep power turns, and the correct entry and recovery procedures for stalls. All maneuvers at critically slow airspeed must be completed no lower than 3,000 feet AGL.

It is very important during this lesson to become familiar with the airplane. Even though it is tough to become comfortable in any airplane in just one lesson, make your strategy to know the airplane systems and layout during this lesson. The lesson should begin with understanding the airplane's systems, especially propeller, electrical, landing gear, and engine instrument systems. Try to arrive at least 30 minutes early and get permission to look around the airplane, outside and inside, without the instructor. No-cost training such as this will eventually pay high safety dividends.

The panel of even the smallest twin-engine airplane will look much more complicated than a single-engine panel. As you look more closely, you will see the duplication of controls and instruments used to adjust and monitor the separate engines. There will be two of everything pertaining to the powerplants: two manifold-pressure needles, two tachometer needles, two throttles, two oil-pressure needles, and the like.

By grouping the paired instruments and indicator needles in your mind, the panel becomes more comfortable and understandable. The "fistful of throttles" simply becomes like one throttle. Make sure you know where the panel and cockpit components are located: engine instruments, electric switches, power instruments, emergency landing gear extension apparatus, and the like. You never want to hunt for something in flight.

Sit in the pilot's seat and close your eyes. Without peeking, point to the right-engine tachometer indication, for instance. Then open your eyes and see if you are indeed pointing correctly. Repeat the session until you can draw the panel in your mind's eye.

During the formal lesson, the instructor should explain that airplane's systems: prop governors, battery-drain vents, battery airflow vents, airplane heater vents, fuel tank vents, stall warning horns, anti-ice or deice equipment, auxiliary power receptacle, and airplane antennas. Ask questions because you need to know about everything attached to or hanging from the airplane.

Settle into the cockpit and start the engines according to the airplane's checklist. Everything will have a specific order, so be careful not to inadvertently skip any items. The left engine is generally considered the number one engine, but there might be some particular reason why you should start the right engine first. It is a good idea to start the engine that is closer to the battery so that the current flow would have least resistance. If the start order is not specified by the manufacturer, some operators suggest that you alternate the engine started first from one flight to the next. This would keep the time on the engines even.

When the engines are started, check all systems for green, and it will be time to learn to taxi. The combined steerable nosewheel and main gear brakes are not new, but multiengine airplanes are bigger and heavier, and are therefore harder to handle in tight places.

Holding brakes and riding brakes while taxiing is never a good idea, so multiengine pilots must become accustomed to using *differential power* for turns. In a tight right turn, the right-engine power can be brought back to idle while the left-engine power is brought forward. The left engine pulls the airplane around (Fig. 7-1).

You must become completely comfortable with taxiing the airplane. One pilot took a multiengine checkride and failed because he could not park the airplane at the end of the ride. He successfully passed the oral exam and the flight portion. The parking space that was available to the applicant required a tight turn. The examiner allowed five tries to get the airplane into position, but eventually the examiner's patience ran out.

The airplane had always been parked in that same tight spot. The flight instructor did not want the student to damage the airplane; therefore, the instructor had always parked the plane at the end of each lesson. The student had never been allowed to park; he never had the opportunity to learn.

On the retest of the multiengine applicant, the examiner had the applicant pull out of the space, taxi around the ramp like an obstacle course, and park the airplane in the original spot. The student passed. The moral of the story: Learn to taxi the airplane, do not park in extremely tight spaces and, when necessary, shut down the airplane and get a tow bar.

The pretakeoff checks and runup should be done slowly and carefully. Follow the checklist exactly. Ask questions when they come to mind. Try to perform the runup in a location that does not block other pilots. Do not feel pressured to finish the checklist in a hurry and, for goodness' sake, do not become a Hobbs-meter watcher here.

Do not feel slighted if the instructor says, "If anything happens on this first takeoff, I have it." The first takeoff should be normal, so if anything abnormal happens, the instructor will handle it. You will have plenty of chances to deal with emergencies on the runway later. Follow instructions and enjoy that first feel of acceleration pushing you back in your seat.

In the air, you will be pleasantly surprised at how easy the airplane is to handle. It will feel heavy at first, but your touch will develop quickly. It will take some time to synchro-

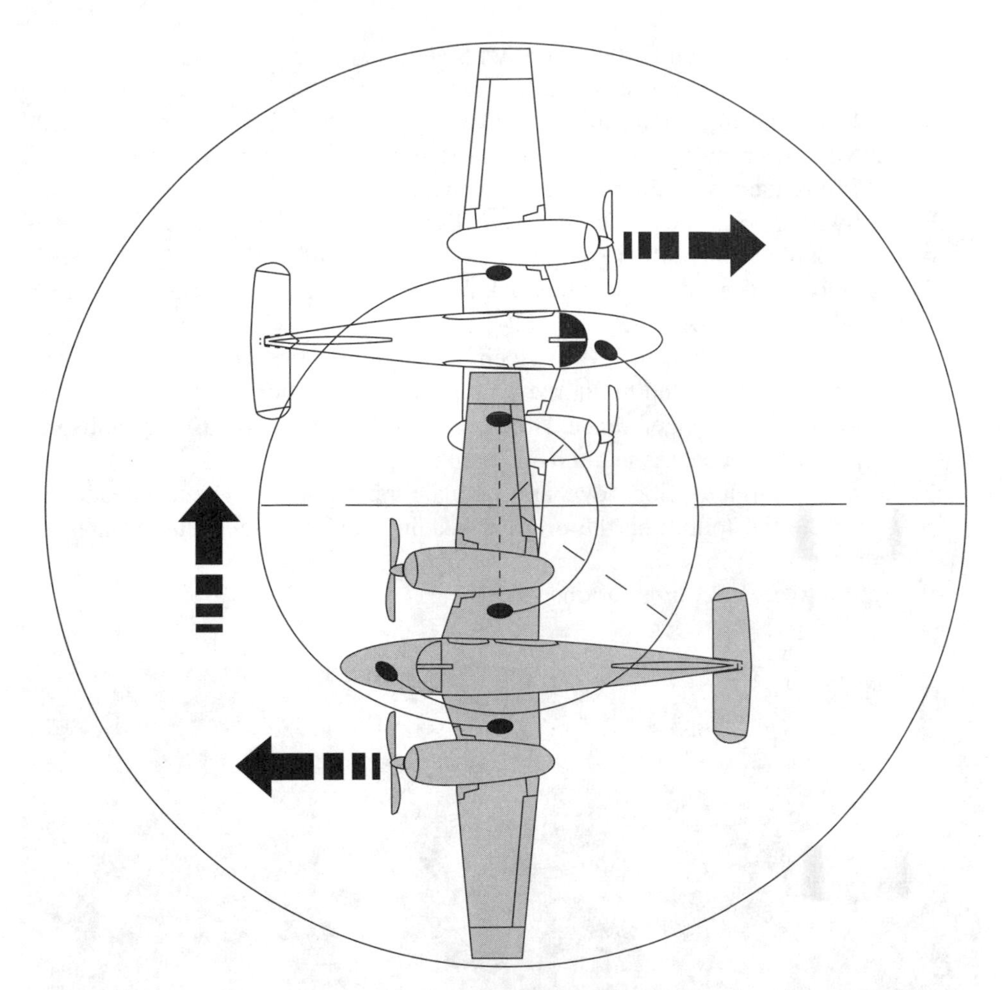

Fig. 7-1. *In a tight spot, the engine on the outside of the turn can be advanced while the inside engine's power remains at idle. This allows the turning radius to become tighter.*

nize the props for the first time. Set one engine and work the prop control to match the RPMs. This takes a little ear training, but it comes with practice. Multiengine training is all business, but initially the instructor should let you fly the airplane to get a very good feel for it without any specific maneuvers. Enjoy the familiarity segment, then get to work.

Steep power-turns

Steep power-turns display airmanship and knowledge of the airplane's characteristics. It is essentially the same maneuver that you have performed in a single-engine airplane (Fig. 7-2):

1. Set the power so that the speed of the airplane is less than V_A.
2. Clear the area with either one 180° turn, or two 90° turns in opposite directions. Make sure that you really look for traffic in the turns.

3. Perform a clean-configuration GUMPS check.
   a. Gas from both engines ON.
   b. Undercarriage UP for this maneuver.
   c. Mixtures rich (or appropriate lean setting).
   d. Props at cruise setting for this maneuver.
   e. Switches or systems (fuel pumps) ON.
4. Begin a roll, either left or right, to a 50°–55° angle of bank.
5. Anticipate the need for elevator back pressure, and raise the nose as necessary while the bank steepens.
6. Stay in the turn through 720° (two circles). While in the turn, observe the outside horizon and keep the horizon in a constant position relative to the dash board as your pitch reference. Hold pitch and bank constant after establishing the turn; do not chase the pitch and bank.
7. Start the rollout to wings level approximately 20° prior to the entry heading.
8. Complete the rollout on the original heading, and maintain altitude and airspeed.
9. Return to cruise configuration.

Fig. 7-2. *The flight instruments during a steep left turn. The bank angle is 50°–55°, the vertical speed indicator is showing a level turn, the ball is in the center, and there is sufficient airspeed above accelerated stall.*

Slow flight

One excellent way to get the feel of an airplane is airspeed transition maneuvers. Maintain straight-and-level flight during airspeed changes will make you adjust the pitch and power to new positions. This forces you to feel how slight changes affect the airplane. This fine-tunes your touch to the airplane. The following is the procedure I use to establish the slow-flight maneuver. Double-check your airplane manufacturer's recommendations:

1. Maintain a safe altitude (at least 3,000 AGL) and an assigned heading.
2. Reduce power on both engines to 15" manifold pressure.
3. While slowing down execute a clearing turn, either one 180° turn or two 90° turns in opposite directions.
4. As the speed continues to slow from cruise, perform the GUMPS check for a landing configuration:
   a. Gas is turned ON for both engines.
   b. Undercarriage DOWN when the speed of the airplane is slower than V_{LE} (maximum landing gear extended speed).
   c. Mixtures to full rich (or properly leaned for the conditions).
   d. Props forward to high RPM.
   e. Switches. Turn fuel pump switches on.
5. Slow the airplane to 70 knots by lowering flaps, one notch at a time, until fully lowered. Hold altitude by adding power as necessary.
6. Hold 70 knots with a constant heading and altitude.
7. Right and left turns can be made while holding 70 knots and a constant altitude.
8. Return to cruise flight speed:
   a. Both throttles to full power.
   b. Flaps up.
   c. Landing gear up when a positive rate of climb is established.
   d. Accelerate to cruise airspeed.

The slow-flight maneuver helps you understand the pitch and power relationships that are required to change speed and maintain altitude. Also, the slow-flight condition is a good platform for other maneuvers.

Approach to landing stall

Aerodynamics of a stalled wing are the same for either single-engine or multiengine airplanes; however, because the engines are on the wings, stall characteristics are altered because the prop wash from the engines alters airflow over wings. Lift is produced as the airflow accelerates over the upper contour of the airfoil. Faster airflow acceleration creates more lift, which means that lift will be greatest where the prop wash is blown across the wings; therefore, power setting and stall condition are closely related.

Air accelerates off the propellers and back across the airfoil and engine nacelle. This accelerated slipstream produces lift that otherwise would not be present. This is one reason why you should never chop power on a multiengine airplane because you are chopping the wing's lift.

The stall recovery in a multiengine airplane is somewhat different than in a single. The stall recovery usually requires a pronounced lowering of the nose to lower the wing's angle of attack from the critical angle. With a multiengine airplane, the nose-down pitch can be shallower if power is added in the recovery. The addition of power, probably full power, will cause high-energy airflow to be sent over the wings. This brute force of prop wash forces the airflow to follow the wing's camber, which produces lift.

A test airfoil in a wind tunnel proved this point about accelerated slipstream and its effect on lift at slow airspeeds. Test data clearly showed that multiengine stall recovery is greatly enhanced with power. Graphs from the wind tunnel test (Figs. 7-3 and 7-4) show that at a 10° angle of attack, with the prop windmilling, the wing produced a coefficient of lift of 1.93. At the same angle, the wing produced a coefficient of 2.88 when the power was on and the airflow accelerated by the propeller. That is a 67-percent increase in lift without changing the pitch angle.

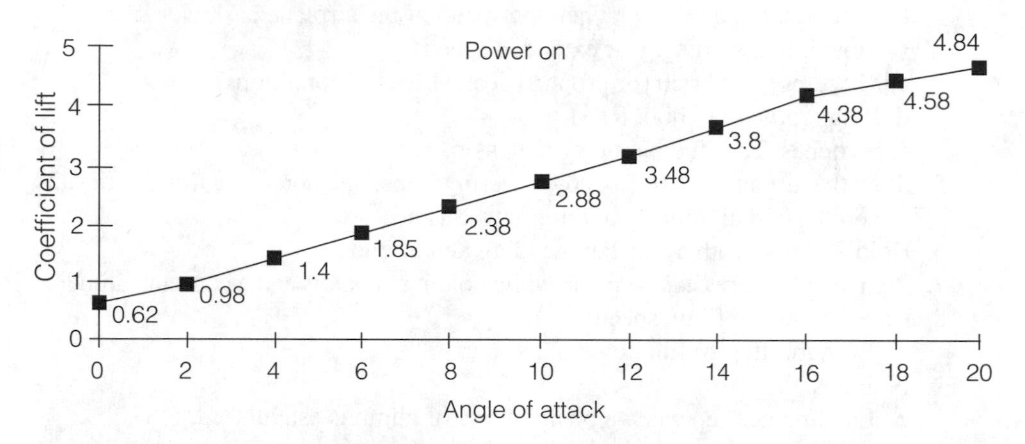

Fig. 7-3. *Wind tunnel test data of lift produced while power is on. The accelerated slipstream from the propeller produces lift as it flows over the engine nacelle and affected wing section.*

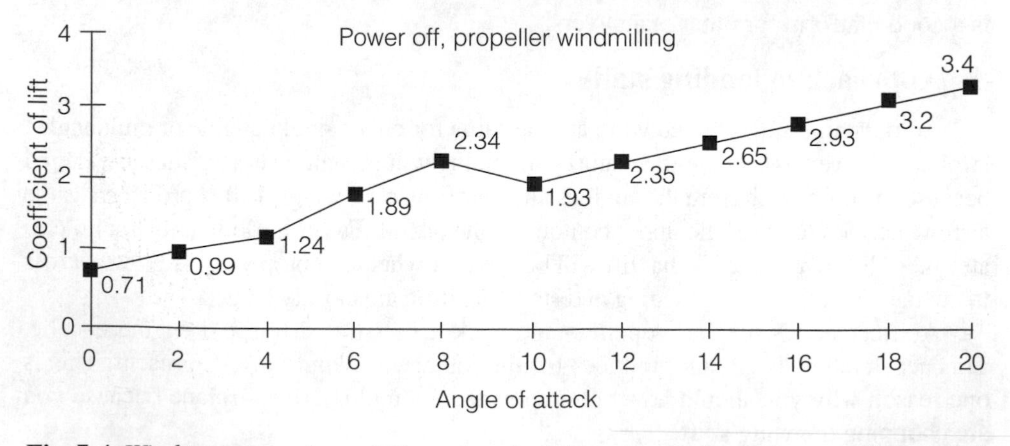

Fig. 7-4. *Wind tunnel test data of lift produced while power is off and the propeller is windmilling. Comparing the chart with Fig. 7-3, the lift advantage provided when the engine has power on is quite clear.*

The effects are even greater when closer to the critical angle of attack. When the test airfoil was placed in the tunnel with a 20° angle of attack, the wing was buffeting wildly and producing a coefficient of 3.4; the engine was turned on and the slipstream created a coefficient of 4.84. This represents a 70 percent increase in lift at the time you need it the most.

This showed that the fastest way to regain lift is engine power in the stall recovery, coupled with a small pitch change. When a single-engine pilot stalls a multiengine airplane for the first time, this fact is sometimes lost, and she will push the nose over too far in recovery. This unnecessarily adds to the altitude loss.

The entire maneuver from start to finish should be accomplished within ±100 feet. Clear the area, and watch for traffic throughout the maneuver. To perform the approach to landing stall maneuver, use the following procedure (Figs. 7-5 and 7-6):

1. Set up the slow-flight platform no lower than 3,000 feet AGL.
   a. 15" manifold pressure.
   b. Clearing turns while reducing speed from cruise.
   c. GUMPS check for landing configuration.
   d. Arrive at 70 knots with full flaps and holding altitude.
2. Reduce power to approximately 13" manifold pressure.
3. Uniformly add pitch to slow the airplane.
4. Hold heading and altitude. The ball of the turn coordinator is in the center.
5. Allow the nose to pitch into a stall attitude. Wait to feel the stall buffet, then go to step 6.

Fig. 7-5. *This photo was taken just prior to a stall. The nose is high, and the vertical speed shows a climb. Note the instrument panel's placard about stalls with only one engine operating.*

Fig. 7-6. *This photo was taken during the stall recovery. The nose is placed on the horizon, the vertical speed shows a descent, and the airspeed is increasing.*

6. Initiate the recovery by simultaneously lowering the nose to the horizon, but not lower than the horizon, and advancing the throttles to full power. Raise the flaps. Retract the landing gear when a positive rate of climb is established.
7. Ease back on the power, and return to normal cruise configuration.

Takeoff and departure stall

Understanding stall characteristics at full power is very important because the deck angle at the time of the stall will be very high due to the accelerated slipstream over the wings. It might not stall until reaching 45° nose-up pitch. This can be a little scary and single-engine pilots are always a little reluctant to pitch that high. You must be careful with any stall; you must be especially careful with a multiengine, power-on stall:

1. 15" manifold pressure. Set up the slow-flight platform no lower than 3,000 AGL.
2. Make clearing turns as the airplane reduces speed.
3. Perform a GUMPS check for a clean configuration.
   a. Gas from both engines ON.
   b. Check that the landing gear is UP and leave it there.
   c. Mixtures rich.
   d. Prop full forward.
   e. Fuel pump switches on.

4. Reduce speed while maintaining altitude until it is time to apply power.
5. Advance power to the desired power setting (21" manifold pressure, climb power, or full power).
6. Raise the nose until the stall occurs. Wait to feel the buffet, then go to step 7.
7. Initiate recovery. Simultaneously place the nose on the horizon and, if power was less than full, add full power.
8. Return to normal cruise configuration of speed, power settings, and prop settings.

Stall tips

Always stall with plenty of recovery altitude. If a cloud deck is at or lower than 3,500 feet AGL, fulfill the ground school portion, and reschedule the flying portion of that lesson. There is a high potential for a spin entry during multiengine stall practice. Regardless of the number of engines, a spin can be entered whenever an uncoordinated stall occurs. An uncoordinated stall occurs when there is yaw present during the stall.

Multiengine airplanes are extremely yaw-prone because the engines do not pull through the center of the airplane. The pilot can never be completely sure that both engines are turning at exactly the same RPM producing exactly the same thrust. When thrust is uneven, the airplane will be uncoordinated. As the pilot raises the nose to reach a stall angle, the engines are turning, but probably not exactly in unison. For this reason, multiengine airplanes are prone to spin.

The recovery from a spin entry in a multiengine airplane will require at least 2,000 feet to initially recognize the spin and pull out. The exact altitude loss is unknown because the FAA does not require multiengine airplanes to be spun for airworthiness certification. This means that if you enter a spin in a multiengine airplane, you become your own test pilot.

Never stall a multiengine airplane while one engine is feathered or while power is reduced on one engine. Stall only when the power settings are as close to symmetrical as possible. Stalls can be performed from straight and level flight or in a turn.

Finally, never look back at the tail, especially a T-tail, when stalling. It is safe, of course, but the tail usually vibrates and sways. This alone can scare you away from stalls.

Simulated engine failure en route

The first lesson is a good time for the instructor to demonstrate en route flying with an engine out. Perhaps the instructor will have you follow through, while she is in complete control of the airplane. While the airplane still has an altitude of at least 3,000 feet AGL, the instructor will reduce the power on one engine, and you can feel the rudder pressure that will be necessary to hold the airplane on a particular heading.

With one engine simulating zero thrust, this is a good time to get a feel for the airplane in various drag configurations. Recall from ground school that small twins simply do not climb well on a single engine. Also recall that the performance charts are based upon ideal conditions; the chart numbers are the absolute best you can hope for.

You will see and feel the 80-percent loss of performance. Clear the area and watch for traffic throughout the drag configuration maneuver:

1. From cruise speed in a clean configuration, reduce power on one engine to idle, and advance power on the other engine to full.
2. Add some power to the idle engine to simulate zero thrust (the effect on the airplane if the propeller were feathered). About 11"–12" manifold pressure will do it.
3. Maintain a speed of V_{YSE} (blue line).
4. Reduce power on the operating engine as necessary to maintain level flight.
5. Lower the landing gear, but attempt to maintain altitude and V_{YSE}.
6. Lower the flaps, one notch at a time, until full flaps are down.
7. Add power on the operating engine as necessary in an attempt to maintain altitude. (Chances are that you will be unable to maintain altitude with one engine, even at full power, with the gear and flaps down. It is a helpless feeling to be holding one throttle at full power; yet the airplane is still sinking. Doing this with plenty of altitude is safe. It will surely make you understand that it is impossible to attempt a climb on one engine when the airplane is heavy and dirty.)
8. Initiate recovery from the drag demonstration.
   a. Add power to the idle engine slowly. (The extra rudder pressure will seemingly melt away as the power becomes symmetrical again.)
   b. Simultaneously add power to both engines until reaching full power.
9. Climb back to the desired altitude and resume cruise configuration.

Normal landings

A multiengine airplane will land the same as a single-engine airplane, but the pattern and approach speeds are faster, and you have to control more inertia. A single-engine airplane does not have to slow down very much to enter a traffic pattern. A multiengine airplane must slow down a noticeable amount to safely enter the traffic pattern. The slow-flight maneuver you performed at altitude becomes very important.

Slow down and get set up in the traffic pattern (Fig. 7-7). Compensate for the speed with a pattern that is somewhat wider than a pattern for a single, but not a great deal wider. Review the airplane's handbook for recommended approach speeds. I like to stay 10 knots faster than the blue line for most of the approach, slowing to blue line on final. I stay at blue line speed until the flare. This means that the flare and float will use up some runway at this speed, but I would rather stay at blue line as long as possible just in case a go-around is needed.

Do not chop the power at any time during the approach. Recall that chopping power chops lift; all available lift is crucial when low and slow. You probably will be required to perform power-on approaches where the power is brought all the way to idle only after touchdown.

You might make several full-stop landings and taxi back for takeoffs. Recall that you have more options and improved safety with runway ahead during a multiengine takeoff. A touch-and-go consumes too much of the safety margin.

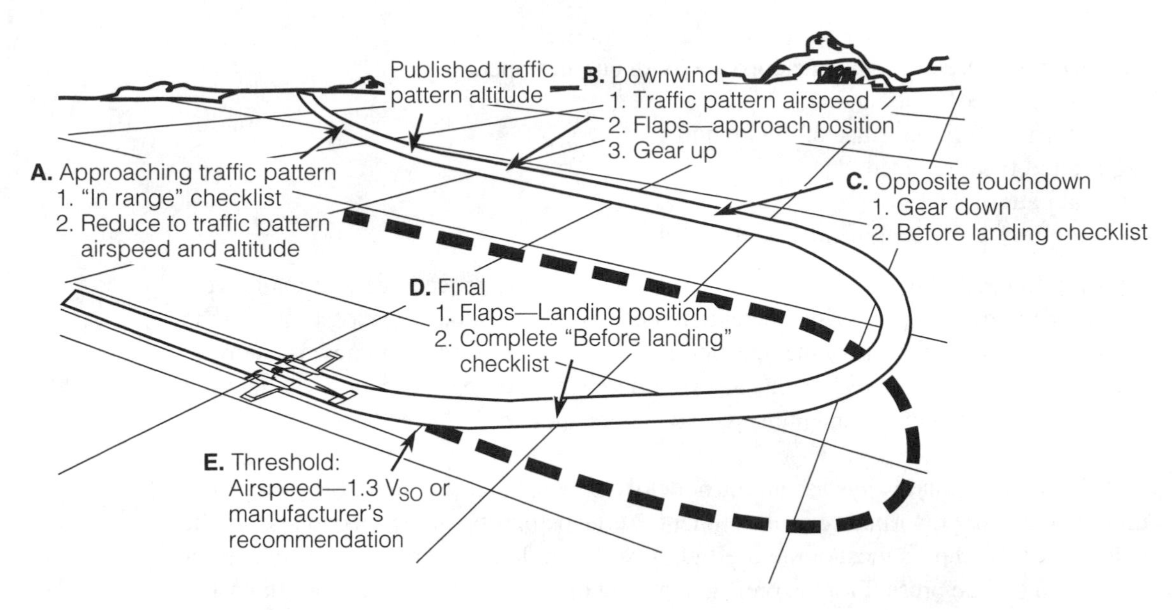

Fig. 7-7. *Normal landing profile.*

After landing

Use the after-landing checklist which is usually an expanded version of "Up, open, and off," which means wing flaps up, cowl flaps open, and fuel pumps off. Prior to engine shutdown, use the checklist to make sure that everything is in order. Stop the engines with the mixture cut-off, not by placing the prop control into feather.

After the engines have stopped and all checklist items are complete, review the flight thoroughly. Resolve any questions that might have come up during the flight. Review the procedures for each maneuver. Plan for the next flight, which should be as soon as possible to facilitate better recall of what you learned this time.

LESSON 2

Objective: The student will practice each of the assigned review maneuvers and procedures to increase proficiency and experience. The student will be introduced to crosswind and maximum-performance takeoff and climb, crosswind and maximum-performance approach and landing, and go-around from a rejected (balked) landing.

Review maneuvers:
Approach to landing stall
Takeoff stall
Steep turns

Introduce:
Maneuvering with one engine at idle

V_{MC} demonstration
Instructor-vectored instrument approach (both engines operating)
Crosswind takeoff and landing
Maximum performance takeoff and landing
Go-around from rejected landing
Normal pattern
One-engine pattern (engine idle on downwind)

Completion standards: The student will perform all the procedures and maneuvers listed for review at a proficiency level that meets or exceeds the criteria set forth in the multiengine-land sections of the appropriate FAA test standards. The new maneuvers and procedures will be evaluated on the adherence to proper procedures, operating techniques, coordination, smoothness, and understanding.

The first lesson always has an introductory atmosphere; the second lesson places more burden for performance on the student. At the beginning of the second lesson, the instructor should pass most of the preflight inspection duties over to you. The instructor should still be present for the preflight, asking the student some questions to ensure comprehension from the first lesson.

Steep turns, slow flight, and the stall series of maneuvers are often grouped together and called "VFR maneuvers." These maneuvers should be practiced at the beginning of the second lesson. The student should now have a good feel for the airplane and be able to perform the VFR maneuvers with precision.

The completion standard for this lesson states that "the procedures and maneuvers listed for review (the VFR maneuvers) (performed) at a proficiency level that meets or exceeds the criteria set forth" in the multiengine practical test standards. This means that steep turns, slow flight, and stalls must be checkride-ready at the conclusion of this lesson.

Maneuvering with one engine

After the VFR maneuvers have been practiced at altitude, it's time to work on single-engine operations again. For this lesson, the engine failure will be simulated by reducing a throttle to the zero thrust power setting. When an engine's power has been reduced to zero thrust, make some turns in both directions. It's OK to turn in the direction of the dead engine; just control your speed.

Many multiengine airplanes have a yaw string attached to the nose. If your airplane does have a yaw string, experiment with the position of the string versus the amount of rudder pressure applied. Use the ball in the inclinometer if no yaw string is used. (Review the yaw string discussion in chapter 1.)

Determine how much rudder pressure will bring the airplane to the zero side-slip position. You will also notice that holding rudder against the engine yaw can be exhausting. You cannot extend your leg and lock your knee; you have to hold muscle pressure continuously while maneuvering. If you can ignore the discomfort in your good-engine leg, the airplane should handle normally.

Demonstration of V_{MC}

Before attempting a V_{MC} demonstration in flight, make sure that you understand what is about to happen and why. Review the V_{MC} discussion on aerodynamics in chapter 1. Ask your instructor plenty of questions. You must know the procedures required by your airplane to get into and out of a V_{MC} demonstration. Do your homework. V_{MC} demonstrations are serious business, and it is no place for an unprepared student.

The demonstration is an experiment to prove that there is a speed where airflow over the rudder is simply not strong enough to overcome the yaw from one wing-mounted engine. This speed changes with altitude. The higher you fly, the less dense the air becomes. When the air is less dense, the operating engine will produce less power. Less power from one side of the airplane means less force from the rudder is required to counteract the engine power from the other side. When less rudder force is needed, less airflow (airspeed) is needed and V_{MC} is reduced.

The horizontal axis of the graph in Fig. 7-8 displays the indicated airspeed of the airplane. The vertical axis is density altitude. Air becomes thinner the farther up the graph you go. The line for V_{MC} is slanted. In this example, V_{MC} is 75 knots at sea level where the air is thick and the good engine is strong. But the line bends backward, showing that V_{MC} is only 40 knots at 7,500 feet.

This information makes it appear that the airplane is much safer at high altitude because you could reduce speed a large amount and still be faster than V_{MC}. It looks safer, but there is a catch; there always is. The indicated stall speed remains the same

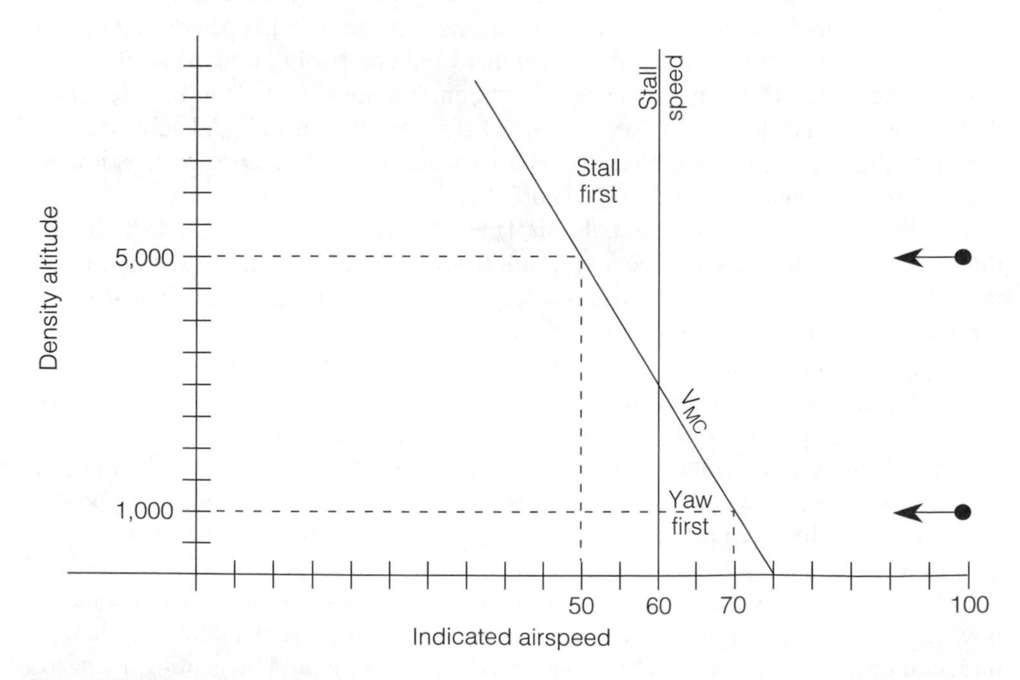

Fig. 7-8. *Density altitude's effect on the indicated airspeed of both stall and V_{MC}.*

regardless of density altitude. The graph shows the stall speed to be a straight vertical line at 60 knots. At approximately 3,000 feet density altitude, the stall line and the V_{MC} line cross. This is a very important position to understand.

Now look at the dot and left arrow at 100 knots and 1,000 feet density altitude. If you and I were preparing to do a V_{MC} demonstration in this airplane, we would first reduce power on the left engine and begin to reduce the airplane's speed. The arrow points in the direction of the speed reduction. We would cross the V_{MC} line at 70 knots, and at this density altitude, the airplane would require a full rudder deflection to keep the airplane straight and under control.

With a 1-knot reduction of airspeed, the rudder force produced would not be strong enough to overcome engine yaw. We would lose control of the airplane if we did not initiate a recovery immediately. We entered the yaw-first region of the graph, and the airplane experienced yaw that we were unable to counteract.

Now we try the same maneuver, but this time at a 5,000-foot density altitude. As we reduce speed from 100 knots with the left engine at idle, we add full power on the right engine and raise the nose. When we arrive at 70 knots, the rudder is still effective, and uncontrollable yaw is not present. This is true because the operating engine is weaker in the thin air at this altitude.

We reduce speed to 60 knots. According to the graph, we still have a 10-knot cushion of airspeed above V_{MC}, but 60 is the stall speed. With the 1-knot reduction in airspeed, the airplane stalls and we never get to V_{MC}. We entered the stall-first region.

When students and instructors go out intending to do V_{MC} demonstrations, they very rarely actually do them. True V_{MC} demonstrations can only happen during conditions with density altitudes that are low, but that would probably place the airplane too close to the terrain to safely do the demonstration. You need altitude to be safe, but a high altitude means that the airplane will stall before reaching true V_{MC}. Unfortunately, the first time a multiengine pilot might ever experience true V_{MC} conditions is low to the ground when an engine quits on takeoff.

During the V_{MC} demonstration, the airplane will yaw or stall, perhaps do both. For this reason, the pilot must initiate a V_{MC}/stall recovery at the first indication of uncontrolled yaw or a stall buffet. To play it safe, a recovery should also be initiated when rudder travel on the good-engine side runs out.

Because asymmetrical power is causing the problem of uncontrolled yaw at any speed slower than V_{MC}, the solution to the problem is to equalize the power again. In demonstration, you can do one of two things. You can advance the power on the simulated bad engine to meet the power setting of the good engine, or reduce power on the good engine to equal the bad engine's power setting. In reality, you must reduce power on the good engine because that is the only option available if you have an actual engine failure.

By reducing the good engine's power to idle, the power output from both sides is now equal. The good news is that you no longer have yaw present, and the airplane is under control. The nose should be down, and the airplane should be gaining airspeed.

The bad news is that you now have no power on either engine and you are dropping like a brick. This is why these demonstrations should never be attempted below 3,000 feet AGL. With enough altitude, the airplane can be accelerated by gravity to a safe speed. Good engine power can be added to stop the descent and maintain level flight.

The left engine should be failed for this demonstration. This leaves you with the right engine, which simulates the most adverse situation where the critical engine is operating. (*See* chapter 1, regarding aerodynamics, for details).

Never attempt a V_{MC} demonstration or a stall with one engine feathered. Although this might be close to realism, it simply does not allow you a safe way out. Practice the V_{MC} demonstration only with an experienced instructor.

Here is the procedure that I teach. Please consult an airplane's operating handbook for its exact V_{MC} demonstration procedure before attempting the maneuver with an instructor onboard:

1. Reduce power on both engines to 15" manifold pressure above 3,000 feet AGL.
2. Make clearing turns.
3. Perform a clean-configuration GUMPS check.
   a. Gas: Fuel selectors ON.
   b. Undercarriage: Verify gear UP for this maneuver.
   c. Mixtures: Rich.
   d. Props: Full forward/high RPM.
   e. Switches: Fuel pumps ON.

4A. Both throttles to full power, then left engine decrease to idle. (Step 4A is more like what would actually happen during an engine failure on takeoff because both engines would be at full power.)

4B. Left engine to idle, then right engine to full power. (Step 4B is more tame and might be preferred for the first demonstration. Either step will yield the desired yaw or stall results.)

5. Establish a 3°–5° bank in the direction of the good engine. Place the ball one-half width out from the center toward the good engine.
6. While the left engine is at idle and the right engine at full power, begin to raise the airplane's nose. Lose approximately 3 knots per second.
7. Use ever-increasing amounts of rudder pressure to maintain directional control.
8. Continue to allow the speed to decrease and the forward travel of the rudder to increase until:
   a. An uncontrollable yaw begins.
   b. A stall occurs.
   c. The rudder pedal touches the floor.
9. At the first indication of yaw, stall, or full rudder travel, initiate recovery:
   a. Reduce the right throttle to a setting where directional control can again be maintained.

b. Lower the airplane's nose and begin to gain airspeed faster than V_{MC} and/or stall.
c. As airspeed increases and directional control is completely regained, add power on the right engine sufficient to stop the descent and achieve straight and level flight at a safe speed of V_{YSE}.

10. Complete the demonstration by smoothly reapplying power to the left engine. Change altitude as necessary. Resume cruise configuration.

Figure 7-9 was taken during the setup to a V_{MC} demonstration. The left throttle is being retarded and the manifold pressure is shown at 18 inches and diminishing. The right throttle is advanced at 25 inches. The vertical speed indicator is seen with a 1,000 fpm climb. This altitude gain will be lost during the recovery. One goal of the maneuver is to recover and stop the resulting descent prior to falling below the altitude used to begin the maneuver. This simulates a climb from the ground, an engine failure, a recovery, and a return to level flight before hitting the ground.

Fig. 7-9. *This photo was taken during the approach to V_{MC}. The left engine power has been reduced. The manifold pressure gauge is visible in the photo. Right engine is set at takeoff power of 25" and the left engine is falling through 18".*

Crosswind takeoff and landing

Crosswind technique is one of the easiest skills to lose. Teaching the crosswind takeoff and landing to student pilots is hard because so many things are happening at once. Many private and commercial pilots are not proficient because crosswind practice might have been rare and abbreviated. Instructor and student must take advantage of any crosswinds that occur during any lesson, which would ensure proficiency.

A crosswind takeoff in a multiengine airplane uses essentially the same technique as a single-engine takeoff. Consult the manufacturer's recommendations for crosswind takeoff and landings. A crosswind component chart is usually included with the airplane's literature. Also check what flap settings are allowed during crosswind operations.

As the takeoff roll begins, the aileron should be turned into the wind. This means that the upwind aileron is deflected up. The upward deflection will hold the upwind wing on the runway. Pilots of multiengine airplanes can use differential power at the beginning of the ground run. By applying more power on the upwind engine, the airplane will resist weathervaning.

The tail is sticking up into the wind. A crosswind strikes the tail on only one side, which tends to pivot the airplane so that the nose aims more into the wind, like a weathervane. If the power is greater on the upwind engine, the differential power will counteract the weathervane force, and the airplane will track straight down the runway centerline.

The rudder will eventually become effective. When the rudder is able to overcome the weathervane force, the pilot should add full power to both engines, and continue the takeoff run as normal. The airplane should be allowed to accelerate through V_{MC} + 5 knots, to a slightly faster than normal rotation speed. This will allow the airplane to pop off the ground. By doing this, the pilot can immediately bank the airplane and establish a crab angle without fear of hitting the upwind main gear or wingtip on the ground. From here, the normal procedures and speed milestones for multiengine takeoff can be followed.

Landing a multiengine airplane in a crosswind is again similar to landing a single-engine airplane in the same conditions. There are two schools of thought.

One method is to hold a crab angle while on final approach and then during the flare to touchdown, apply downwind rudder to take out the crab angle, and align the airplane with the centerline. This requires the pilot to simultaneously apply elevator back pressure and rudder. The elevator and rudder actions require some pilot finesse, but this finesse is hard to come by when both actions happen at the same time.

The second crosswind landing method is to establish a slip on final approach. Aileron holds the upwind-wing down. Opposite rudder prevents the airplane from turning. If the proper amounts of aileron and rudder can be found, the airplane can be held in alignment with the runway. The pilot begins the flare with the upwind-wing low, landing on the upwind main gear first.

The second method does not require the rudder pressure to change during the flare. The pilot should understand what flap settings are allowed while slipping. Avoid the wing-low method when the flaps are deployed beyond the airplane's limit.

No matter which method is used to get on the runway, the aileron should be used during the rollout to hold down the upwind wing. Always use ailerons on the ground for winds. Pull elevator back pressure while taxing over bumps and while braking. Never stop flying the airplane until it is tied down.

Maximum performance takeoff and landing

The maximum performance takeoff is a short-field takeoff. The goal is to get off the runway early and climb over obstructions quickly. A single-engine soft-field technique is not used because a multiengine airplane is limited by V_{MC} as to when it should be allowed into the air. The single-engine technique allows the airplane into the air at the slowest possible speed while utilizing aerodynamic ground effect. The dangers of multiengine flight close to the ground at or slower than V_{MC} outweigh the dangers of a longer ground roll. For this reason, the airplane should still be rotated and flown off the runway at V_{MC} + 5 knots:

1. Hold brakes and add full power to both engines. Verify that full power is being developed, and all engine instruments are indicating in the green arcs.
2. Release brakes and begin ground run. (It might be hard or even impossible to hold the airplane still with brakes against full power in many multiengine airplanes.)
3. Airspeed indicator becomes alive.
4. V_{MC} is passed.
5. V_{MC} + 5 knots. Anticipate rotation.
6. Leave the ground and accelerate to V_X.
7. Retract the landing gear as soon as there is no more runway ahead.
8. Hold the pitch attitude that produces V_X speed until obstructions are clear. (Use 50 feet for practice.)
9. When obstructions are clear, lower the nose, and accelerate to V_Y.
10. Hold the pitch attitude that produces V_Y until 500 feet AGL.
11. Perform the 500-foot check: power back, props back.
12. Above 1,000 feet AGL, establish the best cruise climb speed or best speed for engine cooling.

Normal takeoffs have three segments: ground roll, V_Y climb out, and best cruise climb out. The maximum-performance takeoff has four segments: ground roll, V_X climb, V_Y climb, and best cruise climb. The pilot's skill level must be high enough to maneuver the airplane safely within a close tolerance of airspeed while just above the ground. The maximum-performance takeoff will either display the pilot's airspeed skills or expose the pilot's lack of airspeed skills.

The maximum-performance multiengine landing is a short-field approach and landing. The goal is the same for single-engine and multiengine airplanes: Approach over an obstruction to land on a short runway. The multiengine technique requires a near-constant power setting all the way to flare and touchdown. Do not pass the obstruction and then chop power. Plan the approach to be deliberately high, then reduce power gradually, and add full flaps.

The path over the obstruction and onto the runway should be a slanted straight line. Pilots who are not completely familiar with the maneuver or their airplane routinely lower the flaps to the lowest position too early, or while they are too low. This forces them to add power to make the runway. They arrive at the runway too low and too slow, then chop the power, and flop down.

The better technique is to place the airplane high on final approach and deploy full flaps when the airplane is in a position that requires full flaps. The pilot can then gradually reduce power, controlling the airspeed while gliding to touchdown.

Instrument approach: both engines operating

After completing a practice session of the VFR maneuvers, I want to go back to the airport to work on the landings. Returning to the airport is a great opportunity to execute an instrument approach. The student needs to fly at least one IFR approach when no emergencies are taking place. This will convince the pilot that IFR procedures are the same no matter how many engines you have; only the speeds are different.

Perhaps it would be wise to take some dual instrument instruction in a single-engine airplane to assure proficiency before transitioning instrument skills to a multiengine airplane. The IFR workload of the pilot in a multiengine airplane can be overwhelming.

Operate the throttles as if they were only one throttle. Fly the approach like you did when in a single, except make allowances for the faster multiengine speed. You will be required to fly the approach with precision, communicate with ATC effectively, manage the multiengine cockpit, and display excellent situational awareness.

Lower the landing gear and complete the GUMPS check immediately before the final approach fix on nonprecision approaches. Do it no later than glide slope interception on precision approaches. Adjust power as necessary; control speed at approximately 100 knots for most light twins.

If a missed approach is required, smoothly add full power to both engines and climb away faster than V_{YSE}. The prop controls should already be full forward after the GUMPS check, ready to handle the application of full power. When the climb is established, raise the flaps and landing gear according to the manufacturer's recommendations.

Approach and landing: single-engine

Depending on how long it has taken to reach this point in the second lesson, and the stamina of the student, I usually finish with an engine-out approach and landing. This demonstration should be accomplished with the power of one engine reduced to zero thrust. Never take a multiengine airplane below 3,000 feet with an engine completely shut down. All simulated engine-out situations on the runway, during climbout, or in the traffic pattern should be with fuel selectors on, mixtures rich, and one throttle at idle. (If I ever deliberately shut down an engine above 3,000 feet AGL, and then I am unable to restart it, I treat the situation as an emergency.)

For this demonstration, I retard one throttle to zero thrust while on the downwind leg. The student must be careful not to let the pattern get so wide that an excessive power setting from the good engine is necessary to make the field. Also, the student must play the wind properly with correction angles and allowances for a headwind on either base or final.

The student should be able to gradually reduce power from downwind to touchdown. With every increment of power reduction, the rudder pressure that is needed to overcome yaw becomes less. Eventually, the good engine will be at zero thrust, or nearly so, and no yaw will be present. The touchdown is normal. This maneuver presents no dangerous challenges unless a single-engine go-around is forced. Hopefully, this lesson's single-engine approach and landing can be made with no interruptions.

After the second lesson, I expect the student to be very familiar with the airplane. Airspeed control must be excellent. Their understanding of how the airplane acts with only one engine operating should be established. This lesson should be seen as an early milestone in training because the student will be doing most of the flying starting with Lesson 3.

LESSON 3

Objective: The student will practice the review maneuvers and procedures to maintain or gain proficiency. The student will be introduced to engine-out procedures, learn to identify the inoperative engine and initiate appropriate corrective procedures, and maneuver the airplane with one engine inoperative. The student will demonstrate loss of directional control and the recovery technique with an engine out.

Review maneuvers:
Slow flight
Approach to landing stall
Takeoff and departure stall
V_{MC} demonstration
Engine failures en route
Maneuvering on one engine

Introduce:
Engine failure on takeoff before V_{MC}.
Engine failure during climbout after gear is up.
Identifying an inoperative engine (student should be able to complete all steps unassisted).
In-flight engine shutdown.
Air start.
Simulated vectors to approach (single-engine approach).
Approach to landing with one engine.
Go-around with two engines and one engine.

Completion standards: The student will be able to identify the inoperative engine and use the correct control inputs to maintain straight flight. The student will have a complete and accurate knowledge of the cause, effect, and significance of engine-out minimum control speed (V_{MC}), and recognize the imminent loss of control. All engine-inoperative and loss of directional control demonstrations must be completed no lower than 3,000 feet AGL.

Lesson 3 begins true emergency management training. The concepts learned previously must now be applied correctly and aggressively. The flight instructor was your advocate the first two lessons; now the instructor seemingly becomes your nemesis.

Engine failure

Training and checkride prediction: You will be presented with an engine out on the takeoff roll before reaching V_{MC}. The instructor will present a situation that will require you to think fast and react. Somewhere between "airspeed alive" and V_{MC}, the instructor will say something like, "you have a fire in the right engine," or "you have no oil pressure in the left engine," or another problem that indicates an engine failure is imminent.

The instructor should say which engine, left or right, but it really does not matter. Anytime you are slower than V_{MC}, you cannot fly. That decision is made. The situation is actually presented to see if you will pull the power back on both engines and brake to a stop. During the takeoff run, there is absolutely no time for you to do mechanic work. You cannot, in that split second, try to analyze what is wrong and in which engine.

If you only pull back the "ailing" engine's power, leaving the other engine at full power, the airplane might leave the runway. Do not think about left and right. Any problem during the takeoff run requires immediate power reduction on both engines.

When both engines are at idle, brake as necessary to bring the airplane to a stop without leaving the runway centerline. The instructor might allow you to continue a takeoff run after you have made the correct response to the crisis—if the runway is long enough.

An engine failure after takeoff will be presented. This is sometimes referred to as an "engine cut," but be careful with terminology. An engine-out simulation that is below 3,000 feet should be done using reduced throttle, not by cutting engine power with mixture or a fuel selector valve.

After the airplane is in the air and has passed a position where a landing can be made with the existing runway ahead, the instructor will bring the throttle back to absolute idle. The student must act calmly and correctly without hesitation or confusion. Your primary goal is to maintain control of the airplane by keeping the airspeed faster than V_{MC}. Your second goal is to prevent the airplane from descending into the ground; reduce drag, and fly with maximum available performance.

As soon as the power is reduced on one engine, you will see and feel the airplane sway with the engine yaw. Stop the yaw with rudder. If the nose is moving right, use left rudder. If the nose is yawing left, use right rudder. Use as much rudder as necessary to keep the nose straight.

Now that directional control is maintained, observe the airspeed. Adjust the elevator pitch as necessary to maintain V_{YSE}. Your best hope, maybe your only hope, is to hold that blue line.

Airplane control and speed must be brought under control within the first few seconds of engine failure, then establish best performance:

1. Mixtures full rich.
2. Prop controls full forward.
3. Both throttles full forward. Note: Even though one throttle is dead, you should bring both to full power. The emergency has only been taking place a few seconds. Events are taking place faster than you can see, analyze, and reason. In the first second, you cannot positively identify which engine has failed. The good news is that you do not need to know, just push everything forward. You will be pushing something that is right, even if you do not know which it is yet.
4. Flaps UP.
5. Landing gear UP whenever you see that you are level or in a stabilized climb. Do not bring the landing gear up if you are in a descent and there is still the danger of hitting the ground.
6. Identify which engine has failed. Use the idle foot/idle engine method.
7. Verify which engine has failed by pulling back the throttle of the engine you think has failed. If you are correct, there should be no change in the rudder pressure required to hold the airplane straight.
8. Feather the engine. You will not actually feather the engine in training. When a student calls for the engine to be feathered, I advance the dead engine's throttle from absolute idle to a zero-thrust setting. This will reduce yaw just like feathering the prop would reduce yaw.
9. Continue holding the blueline speed. Climb if possible.

At this point in this lesson, I would give the dead engine back to the student. After full power is regained on both engines, a normal climbout is made. There are some drawbacks to simulating an engine failure by bringing back the throttle, as opposed to using mixture or fuel selectors. The main drawback is that the student can clearly see which engine has been reduced by looking at which throttle has been retarded; the student cannot push both throttles forward. This drawback is not strong enough to overrule the below-3,000-foot-AGL rule. Save the realism for a higher altitude.

The flight instructor will surprise you with a real engine failure at a higher, safer altitude. Even though you know it is coming sometime, it is always a shock. One minute, you are having a friendly conversation; the next minute, the instructor's hand has slipped down toward the fuel selector valve to turn off the fuel to one engine.

The instructor knows that it will take about 30 seconds from the time the valve is moved until the engine actually quits. That is plenty of time for the instructor to continue the pleasant conversation, subsequently making a production out of having her hands nowhere near the mixtures or throttles when the engine actually dies. If the instructor has done the job, your first thought will be that this is a real unexpected failure. Your second thought will be how naive you were to get caught so far off guard.

The position of the fuel selector valve can really help or hinder the stealthy instructor. If the valves are forward under the throttles, the instructor might need to resort to covering the valve panel with the checklist so that you know that something is about to happen but you cannot tell which engine it will be. If the valve is back between the seats, watch out.

I like to turn the valve off and use the 30 seconds before failure to have the student initiate a steep turn. The engine will quit during the turn. While at altitude, I want the student to go through the entire engine-out process. After the initial shock is gone and the student realizes that she is fully responsible to handle the situation, there should be no hesitation:

1. Mixtures rich.
2. Props full forward.
3. Throttles full forward.
4. Flaps up.
5. Gear up.
6. Identify.
7. Verify.
8. Fix or feather. Note: Just above the runway, there is no time to fix the problem, so move quickly to feather the propeller. It is assumed that at altitude, there is time for limited troubleshooting.
9. Decide that altitude allows time to troubleshoot.
    a. Fuel selectors ON. (This must be simulated if the fuel selector valve was deliberately placed in the OFF position for this exercise.)
    b. Primers in and locked.
    c. Carburetor heat ON.
    d. Mixture adjust as necessary.
    e. Fuel quantity check.
    f. Fuel pressure check.
    g. Oil pressure and temperature check.
    h. Magnetos check.
    i. Fuel pumps ON.
    j. Attempt to restart the engine.
10. If the troubleshooting checklist fails to regain engine power, feather and secure the failed engine.
    a. Feather the dead engine's propeller.
    b. Mixture of dead engine to idle cut off.
    c. Dead engine's fuel selector OFF.
    d. Dead engine's fuel pump OFF.
    e. Dead engine's magnetos OFF.
    f. Dead engine's alternator OFF.
    g. Dead engine's cowl flaps closed.
    h. Reduce electrical load as practical.
11. Save the good engine by reducing power as practical and always maintain a speed greater than V_{YSE}.

This is the first time you will be in flight with the engine stopped and the propeller feathered (Figs. 7-10 and 7-11). It is very unusual. The instructor will not let you spend much time flying around while the prop is feathered. For one thing it is not a good idea to leave the hot engine in the icy-cold breeze. It is also not a good idea to overburden the operating engine needlessly. Lastly, the instructor will also be anxious to get the engine going again.

Fig. 7-10. *Maneuvering with one engine inoperative.*

Fig. 7-11. *Feathered right-engine propeller. (This demonstration was conducted on the ramp to safely photograph the feathered propeller up close.)*

For purposes of training the airplane manufacturer probably will include an air-start procedure. The air start might require that the airplane be flown at a particular speed so that the relative wind will help make the propeller windmill. Follow the procedure, and fly the air-start speed. Also, the manufacturer might have specific instructions about the prop control operation and positions.

This procedure does not match every airplane, but it provides some idea about what an air-start checklist requires:

1. Establish air-start airspeed.
2. Fuel selector to the dead engine ON.
3. Dead engine's throttle cracked open about ¼ inch.
4. Dead engine's fuel pump ON.
5. Dead engine's magnetos ON.
6. Dead engine's mixture RICH.
7. Dead engine's prop control full forward.
8. Engage starter to the dead engine. Note: The engine should start to turn, slowly at first, but then faster as the prop comes out of feather. If you have a starter that is temperamental and does not always engage when you start it on the ramp, get that fixed before shutting down that engine in flight. An accumulator that brings the prop out of feather without using the starter is best.
9. After the engine starts, pull prop control back to midrange, approximately 2,000 RPM.
10. Advance the throttle to partial power of about 15 inches. Note: Do not hurry to get all the power back. Allow the engine to run slow until it is warmed up. Keep the cowl flaps closed to retain some engine heat and speed up the warm-up process.
11. Alternator ON.
12. After sufficient warm-up, power can be brought forward to match power on both engines. Perform the cruise checklist.

If the engine cannot be restarted by normal means, the airplane's handbook might have some suggestions about how to get the prop windmilling by alternate means. Of course, flying around with one engine feathered is not a good time to break out the handbook for a little light reading. Look over all the airplane's emergency procedures prior to this flight. If the engine will not restart, and all attempts have failed, proceed below 3,000 feet AGL with extreme caution. Inform the control tower or area traffic of your problem, and this should give you an uninterrupted path to the runway. Be very reluctant to give up the altitude until you absolutely have to.

If an approach must be made with one engine feathered, you must get it right the first time. Single-engine go-arounds will be practiced in your multiengine training, but will be made with one engine at idle not shut down. Actual engine-feathered single-engine go-arounds should be avoided if at all possible; however, the procedure is the same whether the single-engine approach is simulated or for real.

As you approach the runway, be stingy with your altitude. Every single-engine approach has a point of no return (Fig. 7-12). This point is an altitude that might vary with conditions. I usually use 500 feet AGL as my decision altitude. Above 500 feet AGL, if I must make a single-engine go-around, I can (under certain conditions) as long as I immediately go full power with the good engine, reduce drag, and can maintain V_{YSE}.

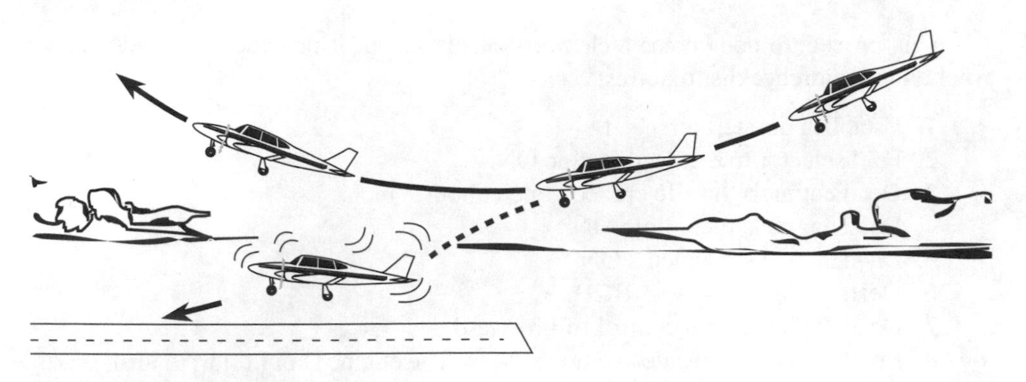

Fig. 7-12. *The decision to go-around on one engine must be made early. There comes a point in each single engine approach beyond which a landing must be made because a go-around becomes extremely hazardous.*

Below 500 AGL, it is probably safer to continue to the ground, even if that means landing in the grass to avoid someone who has pulled out on the runway. Past the point of no return the pilot must not consider going around; the decision to land is made, no matter what else might happen.

During a single-engine approach, leave the landing gear up until the runway is absolutely made. Anticipate the drag and increased rate of descent caused by the landing gear. Use wing flaps sparingly or not at all.

Single-engine instrument approaches

Recall that the greatest hindrance to an IFR multiengine rating is not the multiengine skills; it is the common lack of instrument skills. Many students show up for multiengine training without instrument proficiency and never give it a thought. When I am working with a student toward an IFR/multi, I expect the IFR procedures, skills, and techniques to already be excellent.

If the student is not instrument proficient, the multiengine training will stall and become very expensive. The elapsed time between instrument rating and multiengine rating might have been several years for some pilots. Do yourself and your budget a big favor. Before getting in a multiengine airplane, regain your IFR proficiency in a less expensive single-engine airplane.

The previous lesson had at least one all-engine instrument approach. I try the first single-engine approach during this lesson, if the student is instrument proficient. A simulated engine failure can occur anytime during an approach. The worst time for an engine failure is between the final approach fix and the missed approach point. Under normal circumstances, the airplane is centered on the inbound course, adjusted for wind, and the proper rate of descent has been established. The inbound course and descent can easily be ruined when an engine fails at this position.

Initial yaw caused by engine failure can take you off the course, especially a localizer. The sudden increase in drag will increase the rate of descent, which might cause you to lose the glide slope. Things can go to heck very fast. The pilot must act quickly and aggressively to accurately control two situations at the same time. First, the pilot must meet the engine-failure emergency. Second, the pilot must continue the instrument approach as if everything else is normal.

The most common error is losing track of the approach while reacting to the loss of an engine. Losing track means flying off course, having an excessively steep descent, forgetting the approach timing, or accidentally descending through an MDA or DH. The pilot's workload will skyrocket, so it pays to have prepared early for the approach, which is another aspect of instrument proficiency.

As soon as the pilot gets things under control during a single-engine approach, landing gear and wing flaps must be dealt with. I prefer to perform the complete GUMPS check before beginning the approach so that all those items are taken care of and my attention can be placed on the instrument approach. This means that if an engine fails during the approach, the landing gear and approach flaps are already down.

Bring the gear up? Initially, I would say no, but it does depend on how the approach is expected to terminate. If I am descending on the approach to a straight-in landing, then I will leave the gear down. The landing gear produces drag, which is OK because I want to come down anyway; however, if I expect to fly level at anytime during the approach, I will retract the landing gear. I would try to make the instrument approach terminate with a straight-in landing for this reason of convenience and safety.

LESSON 4

Objective: The student will review the listed maneuvers required for basic instrument flight: maneuvers and procedures, VOR and/or NDB holding procedures, and a VOR approach.

Review maneuvers:
Maneuvering on one engine
Selected VFR airwork maneuvers
Procedures for shut down and feathering
Engine restart
Emergency operations
Systems and equipment malfunctions
VOR approach
NDB approach (The approach will be full or vectored or engine out depending on the student's performance.)
Holding patterns
Missed approach
Single-engine circle-to-land

Completion standards: The student will demonstrate attitude instrument flight while maintaining altitude within 100 feet and headings within 10°. Climbs and descents will be performed within 10 knots of the desired airspeed and level-offs will be completed within 100 feet of the assigned altitude. During engine-out operation the student will be able to readily identify the inoperative engine, shut down, and feather while maintaining altitude within 100 feet and headings with 15°. The approaches will be completed while maintaining the correct approach speed within 15 knots. Holding patterns will be entered correctly and within 10 knots of the proper holding airspeed.

This lesson begins with standard multiengine airwork that should be completely understood. The maneuvers improve proficiency and develop self-confidence. More instrument procedures are practiced after the airwork. I want to get into a holding pattern somewhere and while in the hold I can go through slow flight, and even stalls. Then on to instrument approaches.

The student must handle all the radio communications, fly the airplane with precision, and make all the decisions. In addition to shooting a variety of instrument approaches, I also want to end an approach with a circle-to-land maneuver with only one engine operating.

The previous lesson discussed the decision to keep the landing gear down after engine failure during an approach. Leave the gear down if the landing will be straight in and the path from engine failure to touchdown is a continuous descent. The decision is different if you must fly an extended course while holding a constant altitude during the approach: a circle-to-land maneuver.

The circle-to-land requires the pilot to reach the published approach MDA, fly level around to the other side of the airport, and land. The circling altitude is a mandatory altitude. This is a tricky maneuver under any circumstances, especially with an engine failed.

You cannot legally descend below the circle-to-land minimums until in a position that is within 30° either side of the landing runway's centerline. You cannot climb, or you risk getting back into the clouds, which calls for a missed approach. You cannot go down. You cannot go up. You must maintain straight and level flight with one engine.

Recall the drag demonstrations that showed that a dirty airplane with one engine inoperative might not be capable of flying level. This is a good reason to retract the landing gear in the circle, then reextend the gear when starting final descent to land.

It is also very important to teardrop the approach after a circle-to-land maneuver. Do not try and make a perfect VFR traffic pattern of downwind, base, and final. Play the wind and make the approach one smooth turn to final rolling out on the runway centerline.

Remember that a single-engine go-around from low altitude would be hazardous. Plan the approach properly, and do not overshoot the final approach. If the approach path is faulty, initiate a go-around early. Do not try to salvage a faulty approach with steep-bank turns to realign with the centerline. To do so invites low altitude, unrecoverable stalls.

Plan a point-of-no-return altitude, below which a single-engine go-around is unacceptable. It is possible that your circle-to-land altitude is already at or below your point-of-no-return altitude. This means that when you are committed to a circle-to-land maneuver, you are very close to a full commitment to land.

If you had a choice, you should fly this at an airport with an operating control tower. Inform the tower of the procedure you plan to execute. ATC should be able to keep other airplanes out of the way. At uncontrolled fields, other airplanes might cause an unwanted, and dangerous, single-engine go-around.

A crosswind, partial-panel NDB approach, with an engine failure, concluding with a circle-to-land procedure on the opposite-direction runway, is the supreme test of the pilot's multiengine skills, instrument flying skills, knowledge of the airplane, and good judgment.

LESSON 5

Review and practice:
Fuel flow metering
VOR and/or NDB holding
ILS approach
NDB simulated vector
ILS approach on one engine
VOR and/or approach on one engine (Instructor should try to include a single-engine landing from the approach: straight-in and circle-to-land.)

Completion standards: The student will demonstrate basic attitude instrument flight at a proficiency level that meets or exceeds the criteria specified in the multiengine sections of the applicable FAA test standards. Additionally, the student will demonstrate the ability to perform an approach while maintaining airspeed within 10 knots and altitude within 100 feet on the final approach segment. The student will execute missed approach procedures. If circling approaches are conducted, the student will maneuver the airplane at the MDA in a turn with a radius that does not exceed the visibility minimum for the approach.

Lessons 5, 6, and 7 should all be for proficiency enhancement without much introductory material. The completion standard for Lesson 5 specifically uses the IFR/multi and instrument-rating practical test standards as its benchmark. This means that the student's instrument flying must be checkride-ready from here on. This lesson might take several flights to meet this standard if instrument proficiency was weak to begin with.

The multiengine student should be gaining confidence with each maneuver, approach, and procedure. Pilots at this level are not spectators, but are innovators, and take the lead with decisions. Good judgment should be evident in all actions.

LESSON 6

Objective: The student will practice each maneuver to gain proficiency. The student will learn the procedures required for an engine-out ILS, VOR, and NDB approach.

Review and practice:
All VFR maneuvers
ILS approach on one engine
Nonprecision approach as necessary
V_{MC} demonstration (instrument reference only)
One-engine patterns

Completion standards: The student will demonstrate the ability to perform each maneuver and procedure at a proficiency level that meets or exceeds criteria outlined in the multiengine sections of applicable FAA test standards. After completion of Lesson 6 standards, the student should be ready for the checkride. The last lesson is a mock checkride that is meant to completely prepare the student for the real thing.

LESSON 7

Objective: A chief flight instructor or designated assistant will determine that the instrument-rated student has acquired the proficiency and the performance of the required IFR operations and procedures in the multiengine airplane for successful completion of the FAA flight test.

Review and practice:
All VFR maneuvers
Instrument approach as necessary
Engine failure procedures
One-engine patterns
V_{MC} demonstration
Engine-out taxi

Completion standards: The student will be able to demonstrate each of the listed areas of operations at a proficiency level that meets or exceeds those criteria outlined in the multiengine sections of the applicable FAA test standards. The instrument-rated applicant who desires instrument privileges in the multiengine airplane will be able to demonstrate each of the listed areas of operations at a proficiency level that meets or exceeds those criteria outlined in the instrument-rating test standards.

I have conducted hundreds of these mock checkrides. If I do my job correctly, the lesson will be much tougher than the actual FAA practical test. Lesson 7 should be all encompassing. This means that even if I see something happen that would cause a checkride to be failed, I continue the flight. I want to solve as many potential problems as I can.

On the actual practical test, if the examiner sees something that causes the student to fail, she is supposed to announce it at the time of the failure. The student has an option; stop the test or continue to perform maneuvers after that failure. Either way, the examiner will give credit for the elements of test standards that are passed so that a retest within 60 days applies only to the failed element or elements.

Most applicants are in no mood to continue. The only advantage is to get as much passing credit as possible to shorten the next test session. More often, the test performance after failure goes from bad to worse.

I never say pass or fail during the mock checkride. I continue to test the student even if there are several substandard items. The more problems I can find, the better critique that I can give, and the more help I can give the student in preparing for the actual checkride. That makes this the most demanding lesson of all.

8
Multiengine rating practical test standards

PILOTS NEVER GET PAST "CHECKRIDITIS." THINK OF THE NERVOUSNESS AS part of the test. Applicants blame a checkride failure on the fact that they were nervous, which might not be a valid reason. Plenty of things happen in everyday flying that will make you nervous, in which case you must perform better than any other time.

What if the left engine caught on fire? An engine fire is enough to make anybody nervous. If I am so nervous that I cannot do my job, then I am not a pilot. The problem will get worse if I freeze at the controls and do nothing because I am nervous. Pilots must work through the nervousness and still take corrective action to meet any situation.

What is the worst thing that can happen on a checkride? You can fail. A checkride failure does not scare me as much as an engine failure on takeoff, or getting hit by lightning, or a midair collision. If I stop being a pilot at the first sign of stress and nervousness, then I shouldn't have become a pilot.

Dealing with the checkride nervousness is just as much a part of the test as preflight, stalls, and single-engine landings. We will always be nervous, but it can never be an excuse for our lack of preparation. The best remedy for checkriditis is confidence. Confidence comes from hard work, study, and proficiency.

The checkride will begin with the examiner asking the question, "Is this going to be an IFR or VFR multiengine rating test?" Remember that you are committed when you answer the question. The examiner will look over your logbook and certificate to be sure that you qualify for the test. The examiner will double-check to see that the application for airman certificate or rating form is correctly filled out and signed by the applicant and instructor.

The examiner is supposed to tell you where restrooms are, where the water fountain or soda machines can be found, and whether or not smoking will permitted during the test. After all this, the real test begins.

You will notice that your flight instructor, who has been with you through thick and thin, is nowhere to be found. Applicants always get this great sense of aloneness when the test gets down to business. The first oral exam question is important for psychological reasons. After hours of study, you just want to see if it has all paid off. The effort will pay nice dividends if you have prepared properly.

Do not expect the oral exam questions to be single-information questions. An example of a single-information question: What color is the position light on the airplane's right wing? To answer this question, the applicant has to remember a single piece of knowledge. No reasoning was required, just rote memorization.

The examiner is more likely to ask: "You are flying at night and you see another airplane at your altitude and at your 2 o'clock position. All you can see is a green wingtip light. Who has the right-of-way?" This question requires the applicant to know many more bits of knowledge and to group them together to reason out a proper course of action. In the oral exam expect questions that will force you to produce some judgment.

If you only memorize a list of airplane speeds, capacities, and procedures, without truly understanding how to utilize the speeds, capacities, and procedures, then failure is probable. An examiner might ask single-information questions, but only to set up thought-provoking questions.

The study questions in the next subsection are single-information questions that are included here to help get you started. Many are questions for a chain that starts with a single-information question, then progresses to the real heart of the matter. Go through the questions with the airplane's manuals at your side. The questions concerning a particular airplane can be used as an open-book test of the operating handbook.

MULTIENGINE ORAL EXAMINATION QUESTIONS

Does the airplane have a critical engine?

If the airplane has counterrotating propellers, the answer is no. Only when both engines turn the same direction does one become critical. Refer to chapter 1.

What makes an engine critical?

If both engines turn the same way (clockwise, as seen from the cockpit), P factor will shift the thrust vector to the right. The shifted thrust vector lengthens the lever arm, and yaw becomes more affective. Refer to chapter 1.

What type engine does the airplane have?

Refer to the airplane's operating handbook or other aircraft documents.

What do the numbers and letters of the airplane's engine type mean?

Numbers, such as 360, represent piston displacement. The letter O represents opposed; L represents left-turning or counterrotating. Refer to the airplane's operating handbook or other aircraft documents for specific information.

What is piston displacement?

The piston displacement is the area that all cylinders sweep out during their top-dead-center to bottom-dead-center stroke.

What is the brake horsepower of the engines?

Refer to the airplane's operating handbook or other aircraft documents.

What is the difference between brake horsepower and thrust horsepower?

Brake horsepower (BHP) is the power produced by the engine that is delivered to the propeller. Thrust horsepower (THP) is the actual power from the propeller. Because no propeller is 100 percent efficient, BHP will always be greater than THP. The invention of the constant-speed propeller narrowed the gap between BHP and THP. Refer to chapter 4.

What engine-driven accessories are attached to the engines?

Oil pumps, fuel pumps, vacuum pumps, magnetos, prop governors, tachometers, alternators, and the like are attached to the engines. Accessories and the normal friction of parts inside the engine equal *friction horsepower* (FHP). Engine designers must allow for this loss of power from friction. The theoretical value of horsepower is called *indicated horsepower* (IHP): IHP – FHP = BHP.

How do engine-driven accessories affect brake horsepower?

Because any friction or drag on the engine reduces the power delivered to the propeller shaft, any accessories will reduce BHP. We are willing to accept this BHP reduction in return for the beneficial aspects of accessories: magnetos fire spark plugs; vacuum pumps help spin gyro instruments; and oil pumps move the engine lubricant.

What is the first indication of an engine failure (multiengine airplane)?

When an engine fails, the first indication will be a yaw and roll motion toward the failed engine. The pilot must react instinctively with rudder to keep the airplane straight. Never fly a multiengine airplane with your feet flat on the floor. You must have your feet on the rudder pedals, ready to react.

What items should be systematically checked if engine roughness occurs?

Verify this with the POH, but the primary items are:

- Mixtures—adjust as necessary.
- Fuel selectors—on proper tank.
- Crossfeed—operated properly.
- Electric fuel pumps—on.

- Carburetor heat—on.
- Magnetos—check separately.

How long do you wait for an oil pressure indication before shutting down an engine at engine start?

Verify with the POH, but the rule of thumb is 30 seconds in summer and 60 seconds in winter.

What is the procedure when the cylinder head temperature becomes too hot?

1. Reduce power.
2. Open cowl flaps.
3. Enrichen the mixture.
4. Lower the nose if climbing (best climb cooling speed).

Why is it a poor procedure to idle an engine for long periods of time?

Idling uses the idle circuit of the carburetor, which does not meter fuel as accurately as the main metering system will at faster RPMs. The idle circuit might run the engine rich, which will not burn cleanly. This can cause carbon deposits to foul the spark plugs causing that engine to run rough. It is a good idea to operate the idle circuit (throttle pulled all the way back) for a short period to ensure that the idle circuit is working and that the engine will not quit at low idle.

What is the normal magneto drop during the runup and what are the limitations?

Refer to the airplane's POH. A 100–150 RPM drop is normal, and the difference between the two magnetos on one engine should be within 50 RPM of each other. Allow plenty of warm-up time prior to the engine run-up.

What type of oil should be used in the airplane's engines?

Refer to the airplane's operating handbook or other aircraft documents.

Would a different season of the year or different climate determine which oil to use?

Yes. Outside-air temperatures determine the oil viscosity used in an airplane. Oil must be thick enough to lubricate, but not break down under heat, plus thin enough to properly flow through the engine and lubricate. Check the engine manufacturer's recommendations for proper oil grade within a certain outside-air temperature range.

What is the purpose of carburetor heat?

Carburetor heat provides heated air to the carburetor in the event of icing. Carb heat is also the alternate source of engine air. You should know where the heated air comes from in the airplane. Often the air is heated by contact with a hot exhaust pipe. This always opens the possibility of an exhaust system crack developing and contaminating the carburetor air. You should see an RPM drop when carb heat is applied because the warm air is less dense, which provides less combustion. The RPM reduction proves to the pilot that the heat system is working.

What is the minimum safe amount of oil in each engine?

Refer to the airplane's operating handbook or other aircraft documents.

Can the airplane's oil dipsticks be interchanged?

No. Usually there is a left-engine dipstick and a right-engine dipstick. They do not have duplicate scales. Because the engines are mounted on the wings and because the wings have a dihedral, the oil sits in the pan at an angle. The proper dipstick is needed to get the proper reading of the oil level.

Why are the prop controls pushed forward before landing?

The throttles should be placed full-forward, or at a climb setting in some models, to prepare for a possible go-around. If power were suddenly needed, you would want the propellers to be in position to accept the power. If you inadvertently left the props back, and then added full power, the effect would be like shooting the power of a cannon through the barrel of a rifle. The result would be less than full power (less than full thrust) and possible damage to the engine/prop combination.

What drives the airplane's propeller into a feathered position?

Refer to chapter 4. The airplane's operating handbook and other aircraft documents will have specific information.

How long does it take for the propeller to feather?

Refer to the airplane's operating handbook or other aircraft documents. It should take 8–10 seconds from the time the prop control is placed in the feather position until the propeller actually stops.

What brings the airplane's propeller out of feather?

The airplane's operating handbook or other aircraft documents will have specific information; general information is in chapter 4. Various systems will unfeather propeller blades: a spring in the hub, oil pressure, an accumulator.

What would happen if the prop-governor shaft broke?

If the shaft stopped turning, the flyweights in the governor would no longer have centrifugal force. The flyweights would fold in under the speeder spring's tension. What happens next depends on the airplane's system. Refer to chapter 4.

What would happen if the prop governor's speeder spring broke?

If the speeder spring broke, the flyweights would swing out unchecked by the speeder spring tension. What happens next depends on the airplane's system. Refer to chapter 4.

What is a synchrophaser?

A synchrophaser is the electronic device (described in chapter 4) that automatically places both engines at the exact same RPM. Synchronizing the propellers reduces vibration and the annoying prop noise oscillations.

What holds the gear up in the retracted position?

Refer to the airplane's operating handbook or other aircraft documents. Most are held up by hydraulic pressure.

What happens if the hydraulic system malfunctions or leaks?

Refer to the airplane literature for specific information. Normally, the pressure holding the gear up will be slowly released if the system developed a leak. The landing gear would extend slowly. The pressure sensor of the landing gear system should feel this loss of pressure and bring the gear back up; however, the additional pressure could force the system to leak more. The danger is that if the gear comes out slowly, it will not have the full benefit of a gravity-assist to swing and lock in place. Put the landing gear down while there is still fluid in the system. Land as soon as practical.

How long does landing gear extension or retraction take?

Refer to the airplane's operating handbook or other aircraft documents. It is important to watch for when the landing gear completes its travel. The landing gear's hydraulic pump should automatically turn off when the system pressure reaches its upper pressure limit; however, if the system's pressure sensor malfunctions, the pump will continue to run. This can damage or disable the pump because it is not designed to operate for long periods of time. The attentive pilot should notice if the pump continues to operate past the proper length of time. If the pump does not stop automatically, the pilot should turn off the pump by pulling out the circuit breaker.

What causes the airplane's landing gear to retract?

Most light-twin airplanes use hydraulic fluid pushed by an electric pump. Refer to the airplane's operating handbook or other aircraft documents.

What causes the airplane's landing gear to move to the down-and-locked position?

Refer to the airplane's operating handbook or other aircraft documents.

What can be done if the landing gear does not extend?

The airplane's handbook will have an emergency gear extension checklist. Check the landing gear's electric pump circuit breaker to ensure that it is pushed in. Check the master switch. Verify that the landing gear selector switch is in the down position, then extend the gear manually.

What is the emergency gear-extension speed?

Refer to the airplane's operating handbook or other aircraft documents.

How does the emergency gear extension system work?

Most emergency landing gear extension systems release pressure in the lines, which allows the gear to free fall into place. Refer to the airplane's operating handbook or other aircraft documents for specific information.

What will happen to the landing gear during an electrical failure?

If the system uses an electric pump to move the hydraulic fluid, then the failure of the electrical system will fail the pump. You will need to use the emergency extension system. Refer to the airplane's operating handbook or other aircraft documents.

What is a squat switch and how does it work?

The squat switch is a device that prevents the landing gear from retracting while the airplane is on the ground. The electric landing gear pump will only operate if current is sent from the primary electrical bus, through the squat switch, and then on to the pump. When the weight of the airplane is on the landing gear's oleo strut, the strut is compressed. In this position, the squat switch is open, and the electrical connection to the pump is broken. No electricity gets through; therefore, the pump cannot turn, and the gear cannot fold up. When the airplane is in the air, no weight is on the strut, so the strut extends. The extension closes the squat switch, and the electrical connection is made to the pump. This allows the pump to operate and move the hydraulic fluid, which actuates the landing gear. Refer to the airplane's operating handbook or other aircraft documents for specific information.

Where are any squat switches located on the airplane?

Refer to the airplane's operating handbook or other aircraft documents.

What causes the green gear-down-and-locked light, or lights, to illuminate?

The individual landing gears have a set of contacts, similar to a squat switch, between the primary electrical bus and the green landing-gear-down light. When the landing gear reaches the down-and-locked position, the contacts connect, and the light comes on.

Where is the landing gear hydraulic fluid reservoir located in the airplane?

Refer to the airplane's operating handbook or other aircraft documents.

Why is the landing gear operation speed (V_{LO}) different than the landing gear extended speed (V_{LE})?

Landing gear that is already down and locked can withstand greater loads than during extension or retraction. This is why V_{LE} is faster than V_{LO}. Landing gear doors that are in transit might cup the wind and produce excessive drag, which might adversely affect airspeed at a crucial moment. Depending on the direction of gear travel, an excessive relative wind might thrust the landing gear up into the wells, where they can become jammed.

Where is the brake hydraulic fluid reservoir located in the airplane?

Refer to the airplane's operating handbook or other aircraft documents.

When the landing gear is fully retracted, what makes the landing gear hydraulic system pump stop pumping?

The hydraulic system has a pressure sensor that monitors the number of pounds per square inch that the system is producing in the lines. The pressure sensor has an upper and lower range. When the landing gear is fully retracted, the hydraulic lines will continue to pressurize for a short time after the gear is up. When the pressure builds to the upper pressure limit, the sensor automatically turns off the pump motor. If at any time the pressure falls below the lower limit (leak in a line), the sensor will turn the pump motor on and pump the pressure to the upper limit again. Refer to the airplane's POH to determine the upper and lower pressure limits.

How are the landing gear lights dimmed?

The normal brightness of the landing gear position lights on the panel is too bright for night vision; therefore, the lights need be dimmable. The lights may be dimmed one of two ways. A shutter might be turned to squint out the light. A dimmer setting for all panel lights might include the landing gear position indicators. On some models, the gear lights automatically dim when the navigation lights are turned on. The manufacturer figured that if it is dark enough to turn on the position lights, then it is also dark enough to turn down the cabin lights, including the gear lights. (Students beware. It is a favorite flight-instructor trick to turn on the navigation lights and dim the gear lights when you are not looking. During the daytime, it will look as if the lights are not on at all. Woe be unto you if you land without double-checking the gear lights on short final.)

What is the normal final approach speed?

The safe final approach speed will vary from airplane to airplane. The speed also depends on the use of flaps. Read what the manufacturer has to say about the matter. I like to play it safe and hug the blue line into the flare.

When does the landing gear horn sound?

The landing gear warning horn should sound if the gear handle is ever placed in the UP position while on the ground. If the squat switch system is working properly, the gear should not retract, even with the handle up. The horn will also sound anytime either throttle is retarded below an engine setting sufficient to sustain flight and all three gear are not down and locked, or anytime flaps are extended past an intermediate position while all three gears are not down and locked. Refer to the POH for specific details.

When is the landing gear retracted on takeoff?

The landing gear should only be retracted when no more runway or runway overrun is available. If an engine fails after liftoff, and if there is still enough runway to land straight ahead, you can set the airplane right back down, but only if the landing gear is still down.

What is the recommended main gear and nose gear tire pressure for the airplane?

Refer to the airplane's operating handbook or other aircraft documents.

What does the stall warning horn and landing gear warning horn sound like?

It is very important that you know the difference in the sound of the stall horn and the landing gear unsafe horn. Most manufacturers make the horns sound different; others make one horn a constant sound and the other horn an intermittent sound. Refer to the airplane's operating handbook or other aircraft documents.

The battery is dead, so you call for an APU start. The lineman that brings over the power cart asks, "Do you want this set on 12 or 24 volts?" What will you tell the lineman?

Refer to the airplane's operating handbook or other aircraft documents.

What are the procedures for starting using external power?

The procedure depends on the individual system. Some systems require that the master switch be ON and other have it turned OFF. Refer to the airplane's operating handbook or other aircraft documents.

Where are the batteries located?

Refer to the airplane's operating handbook or other aircraft documents.

Does the battery have ventilation? If so, how is it ventilated? If so, why does it need ventilation?

Vents near a battery will circulate fresh air over the battery. The entrance vent is cut into the wind so that ram-air is forced inside. The exit vent is cut away from the wind so that the air is drawn outside. If the battery ever received an overcharge from the alternator and the battery acid boiled, it would be important to keep the acid's fumes out of the cabin.

Does the airplane have an alternator or a generator?

Refer to the airplane's operating handbook or other aircraft documents.

Which items draw the most electrical current?

The largest electrical draw is usually the landing gear pump motor, followed by the wing flaps' motor, the heater-air blower, the fresh-air blower, the landing lights, and the pitot-tube heater. This information would be important if you found yourself in flight with battery power only. You might need to prioritize the essential equipment based upon power needs.

What happens if the alternator voltage becomes too high?

The system is protected by an over-voltage relay that essentially takes the alternator off line whenever the voltage exceeds a preset limit. The limit is usually 17 volts on 12-volt systems and 31 volts on 24-volt systems.

What is the airplane's maximum alternator output?

Refer to the airplane's operating handbook or other aircraft documents.

What is the difference between an alternator and a generator?

A generator produces electrical power by turning a turbine. An alternator must first receive an electrical current before it can produce a stepped-up current. This distinction is important when it becomes necessary to take an alternator or generator off line due to a malfunction.

What should be done if the alternator overvoltage light illuminates?

When the electrical system warning light comes on, the battery is probably supplying all the electrical power because a spike or other charge in excess of the system tolerance has caused the alternator to be taken off line for its own protection. If the problem was caused by a one-time spike in the system, you can bring the alternator back on line and turn off the warning light by recycling the master switch. If this does not solve the problem, or if the system takes the alternator off line a second time, reduce the electrical load to conserve battery power, and land as soon as practical.

How much fuel does the airplane hold?

Refer to the airplane's operating handbook or other aircraft documents.

One engine has failed en route and has been secured. You decided to extend the airplane's range by transferring fuel from the dead-engine side and burning it in the good engine. What is the procedure for doing this in the airplane?

Refer to chapter 5 for general information. Refer to the airplane's operating handbook or other aircraft documents for specific information.

When should electric fuel pumps be used?

Turn on electric fuel pumps during any crucial phase of flight where an engine failure would be particularly dangerous: takeoff, go-around, switching fuel tanks, and the like. Refer to chapter 5.

Why should electric fuel pumps be used?

Using the electric fuel pumps plus the engine-driven pumps improves the safety margin if an engine-driven fuel pump fails during a crucial phase of flight. Refer to chapter 5.

Where are the electric fuel pumps?

Check the operating handbook for the exact location. You should hear the pumps clicking during engine start.

Would tip tanks affect stall/spin recovery?

Yes. The best spin, if there is such, is nose low, tight, and fast. A nose-low, fast-rotation spin is closer to a recovery attitude than a nose-high, slow-rotation flat spin. Tip tanks that are full of fuel place weight a longer distance away from the center of gravity, and this would cause the spin to fly out and move to a flatter mode. (When figure skaters fly into a spin and tucks their arms in tight, the spin accelerates, but when their arms are extended outward, the spin slows down. Tip tanks are like extended arms.) Tip tanks could induce an unrecoverable flat spin. Remember that multiengine airplanes are not required to do spin testing for certification. If you spin a twin, you become the test pilot.

How are the engines primed?

Refer to the airplane's operating handbook or other aircraft documents.

What is the proper nose strut inflation? What is the proper main gear strut inflation?

Refer to the airplane's operating handbook or other aircraft documents.

What is the ground limitation for pitot heat?

Refer to the airplane's POH for specifics. The pitot heat should be turned on for short intervals while on the ground to prevent overheating. Airflow in flight will prevent overheating.

What is the cold-start procedure?

Refer to the airplane's operating handbook or other aircraft documents.

What is the origin of cabin heat?

Refer to the airplane's operating handbook or other aircraft documents.

Where are the static ports?

The number and location of the outside static ports is important. The static system normally has a port on each side of the airplane. The port lines join to make a Y-shape inside. This Y arrangement reduces static pressure errors due to airplane slip. Avoiding errors in a slip is especially important in multiengine airplanes that due tend to slip during engine out procedures. For the exact port locations, refer to the airplane's POH.

Where is the alternate static port?

Refer to the airplane's operating handbook or other aircraft documents. If the outside static ports ever get clogged by ice or other debris, the pilot must be able to vent the system by using the alternate static source. The pilot must be familiar with the operation and location of the alternate static vent switch.

What are the gyro suction limits?

Refer to the airplane's operating handbook or other aircraft documents.

How is a vacuum pump failure indicated in the airplane?

The suction or instrument air gauge will show a pressure that is below acceptable limits. Refer to the airplane's operating handbook or other aircraft documents.

If the vacuum pump failed in flight, what instruments would be affected?

The vacuum suction or instrument air system turns the gyroscopes in the attitude gyro and the directional gyro instruments. Early detection of a vacuum pump failure is essential so that the pilot can change her scan to include properly operating instruments.

How is the static system drained?

Refer to the airplane's operating handbook or other aircraft documents. Water from rain can easily get into the pitot static system. You have probably seen the three pitot/static system instruments jump while flying in the rain. This happens when water gets into the lines and momentarily causes a false reading. Many systems provide a way to drain the water.

What is the limit on cranking the starter?

Refer to the airplane's operating handbook or other aircraft documents. The starter should never be allowed to grind the propeller around for long periods of time. The maximum time a starter should be engaged is 30 seconds. If the engine does not start, wait at least 2 minutes for the starter to cool before cranking again.

What do cowl flaps do?

Cowl flaps are doors that open underneath the engine compartment. When the doors are open, the relative wind that moves over the opening produces a low-pressure area. The low pressure at the door draws air out of the engine compartment, which aids cooling. Because the cowl flaps do add drag to the airplane, they should be closed when the airplane is flying fast enough that ram air provides adequate cooling. Refer to the airplane's operating handbook or other aircraft documents.

What are the possible flap settings of the airplane?

Refer to the airplane's operating handbook or other aircraft documents.

How are the flaps operated?

Refer to the airplane's operating handbook or other aircraft documents. Some flaps are moved by an electric motor; others are manually operated by a handle.

Where is the external power plug located?

Refer to the airplane's operating handbook or other aircraft documents.

What should be done if a cabin door or luggage compartment door opens in flight?

Return to the field in a normal manner. Have a passenger hold the door during landing flare.

What is the airplane's basic empty weight?

Refer to the airplane's weight and balance form, the airplane's operating handbook, and other aircraft documents.

What is the airplane's maximum landing weight?

Refer to the POH to determine if the airplane has a specific maximum landing weight. The landing weight of many airplanes is less than the takeoff weight. When landing, the airplane is subjected to negative-G loading. The airplane might not absorb negative-G stresses as well as positive G stresses. If the airplane is heavy, and a hard landing is made, the negative-G loading can be exceeded. For this reason, some airplanes must land at a much lighter weight compared to takeoff weight.

Why are maximum ramp weight and maximum takeoff weight different?

Ramp weight is heavier than maximum takeoff weight to allow for fuel burn, which reduces weight during taxi and runup.

Does the airplane have a zero fuel weight?

Refer to chapter 5 and the airplane's POH.

Can the airplane be flown with four average persons on board, full fuel, and full baggage?

Check the weight and balance information carefully. Multiengine training airplanes are not designed for travel comfort. A full load of fuel, passengers, and baggage probably cannot be carried.

What happens when the CG is too far forward?

Rotation and flare are difficult. A pilot risks touching down nosewheel first due to unexpected resistance to the elevator back pressure during takeoff and landing.

What happens when the CG is too far aft?

Rotation might be premature. The airplane's longitudinal stability is reduced, which might provoke a stall. If a stall or spin is encountered, the recovery might be difficult or impossible.

Where is the datum line on the airplane?

Recall that the datum is the reference position for measuring arm to stations. Check the airplane's weight and balance information for the datum position. Many single-engine airplane's use the firewall as the datum. Many multiengine airplanes use the plane of propeller rotation as the datum.

What does the basic empty weight include?

Basic empty weight is the airplane and all permanently attached equipment, including the weight of full engine oil, hydraulic fluids, and unusable fuel. The weight and balance information required to be carried on the airplane will have a list of all items that make up basic empty weight.

What happens to the CG as fuel is burned?

This depends entirely on the position of the various fuel tanks. If the tanks are in the wings, the CG location will not change drastically because the CG range is also within the wing's chord line, but the CG might move slightly. If the CG is already on the forward or aft limit, the range could be exceeded with even a small change in CG caused by fuel burn. Refer to the airplane's operating handbook or other aircraft documents.

What happens to the CG when the landing gear is retracted?

The landing gear folds up and the weight of the tires, struts, and extensions will come to rest in a new position. Depending on the landing gear design, the CG might shift as the gear is retracted or extended. Refer to the airplane's operating handbook or other aircraft documents.

What V speeds make up the fast limit and slow limit of the airplane's normal operating range (green arc)?

The slow end of the green arc is V_{S1} and the fast end is V_{NO}. Refer to the airplane's operating handbook and airspeed indicator.

What V speeds make up the fast limit and slow limit of the airplane's flap operating range (white arc)?

The slow end of the white arc is V_{SO} and the fast end is V_{FE}. Refer to the airplane's operating handbook and airspeed indicator.

What V speeds make up the fast limit and slow limit of the airplane's caution range (yellow arc)?

The slow end of the yellow arc is V_{NO} and the fast end is the redline V_{NE}. Refer to the airplane's operating handbook and airspeed indicator.

Why did the manufacturer of the airspeed indicator leave off any indication of design maneuvering speed (V_A)?

The design maneuvering speed (V_A) changes with the weight of the airplane. V_A increases as the airplane weight increases. Because the airplane is constantly changing weight, V_A cannot be indicated on the airspeed indicator with a constant value.

What does the blue radial line of the airspeed indicator represent?

The blue radial line is only indicated on multiengine airplanes to represent V_{YSE}, which is the best single-engine rate of climb speed. (You will become quite accustomed to hanging on the blueline during multiengine training.)

Why does the multiengine airplane's airspeed indicator have two red lines?

Multiengine airplane airspeed indicators have two red lines, one at the slow end of the speed range and one at the fast end. (Refer to chapter 1.) The slow redline is V_{MC}, the minimum control speed. The fast redline is V_{NE}, the never-exceed speed. Refer to the airplane's operating handbook or airspeed indicator.

What is the maximum demonstrated crosswind component of the airplane?

Most manufacturers include a graph of crosswind components in the airplane's POH. Refer to the airplane's operating handbook or other aircraft documents.

What are the maneuvering limits (positive and negative Gs) of the airplane?

The G loading that an airplane is stressed to handle depends on its certification category. Normal category is stressed to +3.8 Gs and –1.52 Gs. Utility category is +4.4 Gs and –1.76 Gs. Aerobatic category is +6.0 Gs and –3.0 Gs. Refer to the POH or the airworthiness certificate for the category.

How does an increase in altitude affect the true airspeed (TAS) of a stall?

The true airspeed of a stall increases with an increase in altitude.

How does an increase in altitude affect the indicated airspeed (IAS) of a stall?

The indicated airspeed of a stall remains the same as altitude increases.

How does an increase in altitude affect the true airspeed of V_{MC}?

V_{MC}'s true airspeed decreases as altitude increases. Refer to chapters 1 and 7.

What steps must be taken if an engine failure occurs during flight below V_{MC}?

Never let this situation develop. The only recovery is to reduce power on both engines, lower the nose, and accelerate to a speed that is faster than V_{MC} in order to maintain aircraft control. Read more about this in chapters 1 and 2.

When is the gear extended before landing with one engine inoperative?

Lower the landing gear when you are ready to begin the descent from pattern altitude. The landing gear will produce drag, which will increase the sink rate. Use good judgment regarding the wind and other traffic. The gear can be lowered anytime that you need to descend anyway.

What is the final approach speed for the airplane on one engine?

Refer to the airplane's operating handbook or other aircraft documents. I try to hold the blue line until I am sure that a go-around is not needed, then I reduce speed to the manufacturer's recommended speed, followed by flare and touchdown.

When are the flaps extended on a single-engine approach?

Flaps are optional. After the landing gear is down, and the proper adjustments have been made, the flaps can be lowered only after a landing is assured. If the flaps are brought in too early, you might end up struggling to the runway. Adding flaps too early on the approach will cause the airplane to be too low. This will require you to advance power on the good engine. Stepping on the brake and the gas pedal at the same time doesn't make good sense.

How do you crossfeed in the airplane?

Refer to the airplane's operating handbook or other aircraft documents and follow any prescribed checklists. Refer to chapter 5.

What are the normal panel indications when an engine has failed?

The manifold pressure gauge and tachometer will have the same indication as when an engine is operating. The pilot cannot tell by looking outside if the engine has failed because the propeller is windmilling. Rely on the dead-foot/dead-engine method. Read chapter 3. If the airplane is equipped with a fuel-flow meter, a failed engine might be detected by watching a significant reduction in fuel consumption. When the dead engine has had time to cool off, the exhaust gas temperature gauge and the cylinder head temperature gauge will show falling temperatures.

What are the normal panel indications when an engine is shut down?

The engine that has been feathered and secured should have the following engine instrument indications: oil pressure is zero, amp meter is zero, alternator out, vacuum suction is zero, RPM is zero, and manifold pressure reads the ambient pressure.

What is the maximum allowable RPM drop during the runup's feathering check?

Refer to the airplane's operating handbook or other aircraft documents. The norm is a 500 RPM drop.

What are the airplane certification standards for computing V_{MC}?

Refer to chapter 1 for more discussion and details:

- The critical engine is inoperative and windmilling.
- Not more than a 5° bank toward the operative engine.
- Landing gear retracted.
- Flaps in the takeoff position.
- The most rearward CG.
- Cowl flaps in the takeoff position.
- Airplane at gross weight.
- Ground effect negligible.
- Maximum available power on the operating engine.
- Sea-level conditions.
- Trimmed for takeoff.

What is V_{MC}?

Simple definition: The minimum flight speed at which the airplane is directionally controllable with one engine inoperative.

What is the highest single drag item on the airplane?

A windmilling propeller. Have your flight instructor perform a drag demonstration so that you can see and feel the effects of drag on the engine while only one engine is operating. This demonstration is discussed in chapter 7.

What is the second highest drag component on the airplane?

Full flaps. This is why flaps come up early in any recovery procedure and stay up if there is any hope of climbing on one engine. Refer to chapter 3.

How would V_{MC} be affected with a larger engine?

V_{MC} would increase. More power means more yaw that must be overcome with rudder. When greater rudder force is required, the speed must also be faster. Refer to chapter 1.

How would V_{MC} be affected by turbocharging?

A turbocharger makes it possible for an engine to maintain sea-level power at higher altitudes. If power output remains constant, then V_{MC} will remain constant. At a certain altitude, the turbocharger's compressor will be turning at maximum speed, and power will consequently start to decrease as if the engine were normally aspirated. When the power starts to diminish, V_{MC} will get slower. Refer to chapters 1 and 7.

Choose the better takeoff procedure:

A. Rotate at V_{MC} +5 knots, then climb out with a shallow climb angle, and gain as much speed as possible.

B. Rotate at V_{MC} +5 knots, then climb out with a steep climb angle, and hold the blue line.

Answer B is the better selection. Certain pilots might argue that a shallow climbout is easier on passengers, and the extra speed would be helpful if an engine quit. The problem is that the extra speed will quickly evaporate during the first few seconds of the emergency. Altitude is more important. Altitude is extremely hard to come by on just one engine, but altitude gives you the room to adjust airspeed and remain above terrain. The passengers must accept the steeper climbout. Gain altitude as quickly as possible by using the blueline speed of V_{YSE}.

What is accelerate-stop distance?

Refer to the airplane's operating handbook or other aircraft documents. The accelerate-stop distance is the length of runway used by an airplane from a standing start to liftoff speed, and then, after an engine has failed, braking from lift-off speed to a complete stop. If the distance is longer than the runway, the airplane will run off the far end of the runway while the pilot applies the brakes. The distance to stop is always longer than the distance to accelerate to lift-off speed; in certain conditions, the stop distance might be twice as far as the accelerate distance. You should be very familiar with the performance charts pertaining to accelerate-stop distance calculations. Refer to chapter 2.

What is the airplane's single-engine service ceiling and single-engine absolute ceiling?

Refer to the airplane's operating handbook or other aircraft documents. There is a limit to how high any airplane can climb. Airplanes that are not turbocharged are limited to the lower thick-air altitudes. Even with both engines operating, there is a density altitude that is as high as the airplane can go. Recall from chapter 1 that when an engine fails, there is a 50-percent loss of power, but an 80-percent loss of performance. This means that if you are flying at the maximum two-engine altitude when one engine fails, you will find yourself sinking, even though you have full power on the good engine. This is because the single-engine absolute ceiling is lower than the two-engine absolute ceiling. The definition of *absolute ceiling* is the maximum density altitude where the airplane is capable of maintaining level flight. The *single-engine service ceiling* is the density altitude where only a 50-foot per minute climb is possible. Become familiar with the airplane's single-engine service and absolute ceiling numbers. You will be surprised by how low the altitudes are.

If the single-engine absolute ceiling is 4,000 feet density altitude, and the actual density altitude at the airport where you plan to take off is 5,000 feet, should you take off?

The safe answer is no, do not take off until the density altitude has improved. If you do take off and one engine fails, you will not able to climb to safety under any circumstances. The altitude where level flight can be maintained would be 1,000 feet lower than the ground; therefore, any struggle to fly the airplane would be a dangerous waste of time.

You are taking off in a DC-7 (four engines) but the number one engine is inoperative. What will be the three-engine takeoff procedure?

This is the last question. Yes, it is a little unusual. I include it because it is the last question that I was asked on my ATP oral exam. I told the examiner that I had only seen the cutaway nose section of a DC-7 at the National Air and Space Museum. That did not get me off the hook. I went on to explain that flying a four-engine airplane with only three operable engines would take a special flight permit from the FAA. "Yes, of course," the examiner said, "but what about the takeoff procedure?" I said that I would start out with full power on engines 2 and 3, which are the two inboard engines. This would give me symmetrical thrust during the beginning of the takeoff roll. Then, after the speed becomes fast enough to make the rudder effective, I would come in with partial power on the number 4 engine, which is the outboard engine on the right wing. The number 2 and 3 engines would cancel out each other's yaw, and the rudder would cancel out the yaw of the number 4 engine. The examiner accepted the answer. I became an ATP later that day. Good luck to you on the DC-7 questions . . ..

MULTIENGINE PRACTICAL TEST STANDARDS

NOTE: An applicant seeking initial certification as a commercial pilot with an airplane multiengine land class rating will be evaluated in all tasks listed within this section.

An applicant who holds a commercial pilot certificate with an airplane rating other than multiengine land and is seeking the addition of a multiengine land class rating, will NOT be evaluated on those areas of operations/tasks so noted.

I. PREFLIGHT PREPARATION

A. Certificates And Documents
B. Obtaining Weather Information
C. Cross-Country Flight Planning
D. Night Flight Operations
E. Aeromedical Factors

II. MULTIENGINE OPERATION

A. Operation of Airplane Systems
B. Emergency Procedures
C. Determining Performance and Limitations
D. Flight Principles—Engine Inoperative

III. GROUND OPERATIONS

A. Visual Inspection
B. Cockpit Management
C. Starting Engines
D. Taxiing
E. Pre-Takeoff Check

IV. AIRPORT AND TRAFFIC PATTERN OPERATIONS

A. Radio Communications and ATC Light Signals
B. Traffic Pattern Operations
C. Airport and Runway Marking and Lighting

V. TAKEOFFS AND CLIMBS

A. Normal and Crosswind Takeoffs and Climbs
B. Maximum Performance Takeoff and Climb

VI. INSTRUMENT FLIGHT

A. Engine Failure During Straight-And-Level Flight and Turns
B. Instrument Approach—All Engines Operating
C. Instrument Approach—One Engine Inoperative

VII. FLIGHT AT CRITICALLY SLOW AIRSPEEDS

A. Imminent Stalls, Gear Up and Flaps Up
B. Imminent Stalls, Gear Down and Approach Flaps

C. Imminent Stalls, Gear Down and Full Flaps
D. Maneuvering During Slow Flight

VIII. MAXIMUM PERFORMANCE MANEUVERS

A. Steep Power Turns

IX. FLIGHT BY REFERENCE TO GROUND OBJECTS

A. Eights Around Pylons

X. EMERGENCY OPERATIONS

A. Systems and Equipment Malfunctions
B. Maneuvering with One Engine Inoperative
C. Engine Inoperative Loss of Directional Control Demonstration
D. Demonstrating the Effects of Various Airspeeds and Configurations During Engine Inoperative Performance
E. Engine Failure on Takeoff Before V_{MC}
F. Engine Failure After Lift-Off
G. Engine Failure En Route
H. Approach and Landing with an Inoperative Engine

XI. APPROACHES AND LANDINGS

A. Normal and Crosswind Approaches and Landings
B. Go-Around from Rejected (Balked) Landing
C. Maximum Performance Approach and Landing
D. After-Landing Procedures

I. AREA OF OPERATION: PREFLIGHT PREPARATION

NOTE: Evaluation in this area is NOT required for commercial pilot certificate holders with airplane rating who are seeking the addition of an airplane multiengine land class rating.

A. TASK: CERTIFICATES AND DOCUMENTS (AMEL)

PILOT OPERATION - 1

REFERENCES: FAR Parts 43, 61, and 91; AC 61-21, AC 61-23; Pilot's Operating Handbook and FAA-Approved Airplane Flight Manual.

1. Objective. To determine that the applicant:

a. Exhibits commercial pilot knowledge by explaining the appropriate -

(1) pilot certificate privileges and limitations applicable to flights for compensation or hire.
(2) medical certificate, class, and duration.
(3) personal pilot logbook or flight record.

b. Exhibits commercial pilot knowledge by locating and explaining the significance and importance of the -

(1) airworthiness and registration certificates.
(2) operating limitations, handbooks, and manuals.
(3) equipment list.
(4) weight and balance data.
(5) maintenance requirements, tests, and appropriate records applicable to flights for hire, including preventive maintenance and maintenance that can be performed by the pilot.

2. Action. The examiner will:

a. Ask the applicant to present and explain the appropriate pilot and medical certificates and personal flight records applicable to flights for compensation or hire.

b. Ask the applicant to locate and explain the airplane's documents, lists, and other required data including the airplane's maintenance records, and determine that the applicant's performance meets the objective.

B. TASK: OBTAINING WEATHER INFORMATION (AMEL)

PILOT OPERATION - 1

REFERENCES: AC 00-6, AC 00-45, AC 61-21, AC 61-23, AC 61-84.

1. Objective. To determine that the applicant:

a. Exhibits commercial pilot knowledge of aviation weather information including high altitude weather and weather activity over wide geographical areas, by promptly and systematically obtaining, reading, and analyzing -

(1) weather reports and forecasts.
(2) weather charts.
(3) significant weather prognostics.
(4) constant pressure prognostics.
(5) pilot weather reports.
(6) SIGMETs and AIRMETs, including wind-shear reports.
(7) Notices to Airmen.

b. Exhibits commercial pilot knowledge and awareness by explaining aviation weather hazards.

c. Uses critical judgment in making a competent go/no-go decision based on the weather information.

2. Action. The examiner will:

a. Determine that the applicant promptly and systematically obtains all

pertinent weather information. (If current weather materials are not available, the examiner will furnish samples for use.)
b. Ask the applicant to analyze and explain the weather data and aviation weather hazards, and determine that the applicant's performance meets the objective.

C. TASK. CROSS-COUNTRY FLIGHT PLANNING (AMEL)

PILOT OPERATION - 1, 5

REFERENCES: AC 61-21, AC 61-23, AC 61-84.

NOTE: Examiners will relate the required applicant knowledge in this TASK to the most complex airplane used for the practical test. In-flight demonstra tions of cross-country procedures by the applicant will not be required.

1. Objective. To determine that the applicant:

a. Exhibits commercial pilot knowledge by promptly and systematically planning a VFR cross-country flight near the maximum range of the airplane, considering payload and fuel including one leg for night operations.
b. Selects and uses current and appropriate aeronautical charts.
c. Plots a course for the intended route of flight, including fuel stops, available alternates, and suitable course of action for various situations.
d. Selects prominent en route checkpoints.
e. Selects most favorable altitudes or flight levels, considering weather conditions and equipment capabilities.
f. Computes flight time, headings, and fuel requirements.
g. Selects appropriate radio aids for navigation and communications.
h. Identifies airspace, obstruction(s), and terrain features.
i. Extracts and records pertinent information from Airport/Facility Directory and other flight publications, including NOTAM and airport information.
j. Completes a navigation log.
k. Completes and simulates filing a VFR flight plan.

2. Action. The examiner will:

a. Ask the applicant to plan, within 30 minutes, a VFR cross-country flight near the maximum range of the airplane considering payload and fuel.
b. Ask the applicant to explain cross-country flight planning procedures, and determine that the applicant's performance meets the objective.

D. TASK: NIGHT FLIGHT OPERATIONS (AMEL)

PILOT OPERATION - 1

REFERENCES: AC 61-21, AC 67-2.

NOTE: The examiner will evaluate the applicant's knowledge of night flying operations through oral testing only.

1. Objective. To determine that the applicant:
   a. Exhibits commercial pilot knowledge by explaining night visual perception including -
      (1) function of various parts of the eye essential for night vision.
      (2) adaptation of the eye to changing light conditions.
      (3) correct use of the eye to accommodate changing light conditions.
      (4) coping with illusions created by various light conditions.
      (5) effects of pilot's physical condition on visual perception.
      (6) aids for increasing vision effectiveness.
   b. Exhibits commercial pilot knowledge by explaining personal equipment recommended for night flight operations including -
      (1) types and use of various lighting.
      (2) arrangement of equipment.
   c. Exhibits commercial pilot knowledge by explaining airplane lighting and equipment for night flight operations including -
      (1) required equipment.
      (2) additional equipment recommended.
      (3) external light interpretation.
   d. Exhibits commercial pilot knowledge by explaining airport and navigation lighting including -
      (1) meaning of various lights.
      (2) determining status of lights.
      (3) airborne activation of runway lights.
   e. Exhibits commercial pilot knowledge by explaining airplane night operations including -
      (1) preparation and preflight.
      (2) starting, taxiing, and run-up.
      (3) takeoff and departure.
      (4) orientation and navigation.
      (5) night emergencies.
      (6) approaches and landings.
2. Action. The examiner will ask the applicant to explain night visual perception, personal equipment, airplane lighting and equipment, airport and navigation lighting, and airplane night operation, and determine that the applicant's knowledge meets the objective.

E. TASK: AEROMEDICAL FACTORS (AMEL)

PILOT OPERATION - 1

REFERENCES: AC 61-21, AC 67-2.

1. Objective. To determine that the applicant:

   a. Exhibits commercial pilot knowledge of the elements related to aeromedical factors including -

      (1) hypoxia, its symptoms, effects, and corrective action.
      (2) hyperventilation, its symptoms, effects, and corrective action.
      (3) middle ear and sinus problems, their causes, effects, and corrective action.
      (4) spatial disorientation, its causes, effects, and corrective action.
      (5) motion sickness, its causes, effects, and corrective action.
      (6) the effects of alcohol and drugs, and their relationship to safety.
      (7) carbon monoxide poisoning, its symptoms, effects, and corrective action.

   b. Exhibits commercial pilot knowledge of nitrogen excesses during scuba dives, and how this affects a pilot during flight.

2. Action: The examiner will determine that the applicant's performance meets the objective.

II. AREA OF OPERATION: MULTIENGINE OPERATION

NOTE: Because elements of aeronautical knowledge important for safe multiengine operation may not have been previously demonstrated, all items contained in this area of operation that are applicable to the airplane used will be evaluated through oral questioning.

A. TASK: OPERATION OF AIRPLANE SYSTEMS (AMEL)

PILOT OPERATION - 1

REFERENCES: AC 61-21; Pilot's Operating Handbook and FAA-Approved Airplane Flight Manual.

1. Objective. To determine that the applicant exhibits commercial pilot knowledge by accurately explaining the applicable normal operating procedures and limitations of the airplane's systems using correct terminology in identifying components, including:

   a. Primary flight controls and trim.
   b. Wing flaps, leading edge devices, and spoilers.
   c. Pitot static system and associated flight instruments.

d. Vacuum system and associated flight instruments.

e. Landing gear -

(1) retraction system.
(2) indicators.
(3) brakes and tires.
(4) nosewheel steering.

f. Powerplant -

(1) controls and indicators.
(2) induction.
(3) carburetion and fuel injection.
(4) exhaust and turbocharging.
(5) cooling.
(6) fire detection.

g. Propellers -

(1) type.
(2) controls.
(3) feather, unfeather, autofeather, and negative torque sensing.
(4) synchronizing, synchrophasing.

h. Fuel system -

(1) capacity, pumps, controls, and indicators.
(2) crossfeed and transfer.
(3) fueling procedures.
(4) approved grade, color, and additives.
(5) drain valves.
(6) low-level warning.

i. Oil system -

(1) capacity.
(2) grade.
(3) indicators.

j. Hydraulic system -

(1) controls and indicators.
(2) pumps and regulators.

k. Electrical system -

(1) controls and indicators.
(2) alternators or generators.
(3) battery, auxiliary power unit.
(4) circuit protection.

(5) external and internal lighting.
(6) associated flight instruments.

l. Environmental system -

(1) heating.
(2) cooling and ventilation.
(3) controls and indicators.
(4) oxygen and pressurization.

m. Ice prevention and elimination.
n. Avionics.

2. Action. The examiner will ask the applicant to explain the normal operating procedures and limitations of the airplane systems using correct terminology in identifying components.

B. TASK. EMERGENCY PROCEDURES (AMEL)

PILOT OPERATION - 6

REFERENCES: AC 61-21; Pilot's Operating Handbook and FAA-Approved Airplane Flight Manual.

NOTE: Demonstration of intentional spins and recovery are not required on the practical test and are prohibited in many airplanes. However, the examiner will ask the applicant to explain the recommended spin recovery procedure for the particular airplane used. This knowledge is essential for recovery if an unintentional spin occurs. This is a knowledge requirement ONLY. It is not intended that spins be practiced.

1. Objective. To determine that the applicant exhibits commercial pilot knowledge by correctly explaining the applicable emergency procedures including:

a. Emergency checklist.
b. Partial power loss.
c. Engine failure -

(1) before lift-off.
(2) after lift-off.
(3) during climb and cruise.
(4) engine securing.
(5) restart.

d. Single-engine operation -

(1) approach and landing.
(2) restart.

e. Emergency landing -

(1) precautionary.
(2) without power.
(3) ditching.
(4) Use of approved flotation gear and pyrotechnic signalling devices.

f. Engine roughness or overheat.
g. Loss of oil pressure.
h. Smoke and fire -

(1) engine.
(2) cabin.
(3) electrical.
(4) environmental.

i. Icing -

(1) airframe.
(2) powerplant.

j. Crossfeed.
k. Pressurization.
l. Emergency descent.
m. Pitot static system and associated instruments.
n. Vacuum system and associated instruments.
o. Electrical.
p. Landing gear.
q. Wing flaps (asymmetrical position).
r. Inadvertent door opening.
s. Emergency exits.

2. Action. The examiner will ask the applicant to explain the emergency procedures, and determine that the applicant's performance meets the objective.

C. TASK: DETERMINING PERFORMANCE AND LIMITATIONS (AMEL)

PILOT OPERATION - 1

REFERENCES: AC 61-21, AC 61-23, AC 61-84, AC 91-23; Pilot's Operating Handbook and FAA-Approved Airplane Flight Manual.

1. Objective. To determine that the applicant:

a. Exhibits commercial pilot knowledge by explaining performance and limitations including a thorough knowledge of the adverse effects of exceeding the limits.
b. Demonstrates proficient use of the appropriate performance charts, tables, and data including cruise control, range, and endurance.

c. Determines the airplane performance, considering the effects of various conditions, in all phases of flight including -

(1) accelerate-stop distance.
(2) accelerate-go distance.
(3) takeoff performance, all engines, single engine.
(4) climb performance, all engines, single engine.
(5) service ceiling, all engines, single engine.
(6) cruise performance.
(7) fuel consumption, range, endurance.
(8) descent performance.
(9) go-around from rejected landings.
(10) landing distance.

d. Describes the effects of seasonal and atmospheric conditions on the airplane performance.
e. Computes weight and balance, including adding, removing, and shifting weight; and determines if the weight and center of gravity will remain within limits during all phases of flight.
f. Uses mature judgment in making a competent decision on whether the required performance is within the airplane capability and operating limitations.

2. Action. The examiner will:

a. Ask the applicant to explain the airplane performance and limitations including adverse effects of exceeding the limits.
b. Ask the applicant to determine the airplane performance and limitations and to describe the effects of seasonal and various atmospheric conditions on the airplane operation, and determine that the applicant's performance meets the objective.

D. TASK: FLIGHT PRINCIPLES—ENGINE INOPERATIVE (AMEL)

PILOT OPERATION - 6

REFERENCES: AC 61-21; Pilot's Operating Handbook and FAA-Approved Airplane Flight Manual.

1. Objective. To determine that the applicant exhibits commercial pilot knowledge by explaining the flight principles related to operation with an engine inoperative including:

a. Factors affecting single-engine flight -

(1) density altitude-
(2) drag reduction (propeller, gear, and flaps).
(3) airspeed (V_{SSE} V_{XSE}, V_{YSE}).
(4) attitude (pitch: bank coordination).

(5) weight and center of gravity.
(6) critical engine.

b. Directional control -

(1) reasons for loss of directional control.
(2) reasons for variations in V_{MC}.
(3) indications of approaching loss of directional control.
(4) safe recovery procedure if directional control is lost.
(5) V_{MC} in relation to stall speed.
(6) whether an engine inoperative loss of directional control demonstration can be safely accomplished in flight.

c. Takeoff emergencies -

(1) takeoff planning.
(2) decisions after engine failure.
(3) single-engine operation.

2. Action. The examiner will ask the applicant to explain engine inoperative flight principles, and determine that the applicant's performance meets the objective.

III. AREA OF OPERATION: GROUND OPERATIONS

A. TASK: VISUAL INSPECTION (AMEL)

PILOT OPERATION - 1

REFERENCES: AC 61-21; Pilot's Operating Handbook and FAA-Approved Airplane Flight Manual.

1. Objective. To determine that the applicant:

a. Exhibits commercial pilot knowledge of airplane visual inspection by explaining the reasons for the inspection, what items should be inspected, and how to detect possible defects.
b. Inspects the airplane by systematically following an appropriate checklist.
c. Verifies that the airplane is in condition for safe flight emphasizing -

(1) fuel quantity, grade, and type.
(2) fuel contamination safeguards.
(3) fuel tank venting.
(4) oil quantity, grade, and type.
(5) fuel, oil, and hydraulic leaks.
(6) oxygen supply, if appropriate.
(7) flight controls.
(8) structural damage including exhaust system.
(9) tiedown, control lock, and wheel chock removal.

(10) lighting.
(11) ice and frost removal.
(12) security of baggage, cargo, and equipment.

d. Demonstrates proper management of the fuel system.
e. Notes any discrepancy and accurately judges whether the airplane is safe for flight or requires maintenance.

2. Action. The examiner will:

a. Ask the applicant to explain the reasons for the inspection, what items should be inspected, and how to detect possible defects.
b. Observe the applicant's visual inspection procedure, and determine that the applicant's performance meets the objective.

B. TASK: COCKPIT MANAGEMENT (AMEL)

PILOT OPERATION - 1

REFERENCE: AC 61-21.

1. Objective. To determine that the applicant:

a. Exhibits commercial pilot knowledge of cockpit management by explaining efficient cockpit management procedures, securing cargo, and related safety factors.
b. Organizes and arranges material and equipment in a manner that makes them readily available.
c. Adjusts and locks the rudder pedals and pilot's seat to a safe position and assures full control movement.
d. Ensures that safety belts and shoulder harnesses are fastened.
e. Briefs occupants on the use of safety belts and emergency procedures including the use of flotation gear and pyrotechnic signalling device, when aboard.

2. Action. The examiner will:

a. Ask the applicant to explain the efficient procedure for good cockpit management and related safety factors.
b. Observe the applicant's cockpit management procedures, and determine that the applicant's performance meets the objective.

C. TASK: STARTING ENGINES (AMEL)

PILOT OPERATION - 1

REFERENCES: AC 61-21, AC 61-23, AC 91-13, AC 91-55; Pilot's Operating Handbook and FAA-Approved Airplane Flight Manual.

1. Objective. To determine that the applicant:

   a. Exhibits commercial pilot knowledge by explaining correct engine starting procedures including the use of an external power source, starting under various atmospheric conditions, and the effects of using incorrect starting procedures.
   b. Performs all items by systematically following the before-starting and starting checklist.
   c. Demonstrates competence in the care and use of equipment.
   d. Accomplishes correct starting procedure with emphasis on -

      (1) positioning the airplane to avoid creating hazards.
      (2) determining that the area is clear.
      (3) adjusting the engine controls.
      (4) setting the brakes.
      (5) preventing airplane's movement after engine start.
      (6) avoiding excessive engine RPM and temperatures.
      (7) checking engine instruments after engine start.

2. Action. The examiner will:

   a. Ask the applicant to explain correct engine starting procedures and the effects of using incorrect procedures.
   b. Observe the applicant's engine starting procedures, and determine that the applicant's performance meets the objective.

D. TASK: TAXIING (AMEL)

PILOT OPERATION - 1

REFERENCE: AC 61-21.

1. Objective. To determine that the applicant:

   a. Exhibits commercial pilot knowledge by explaining all aspects of safe taxi procedures including the effect of wind on the airplane during taxiing.
   b. Follows the prescribed taxi check list, if pertinent.
   c. Performs a brake check immediately after the airplane begins movement, and thereafter uses proper braking technique.
   d. Complies with markings, signals and clearances, and follows the proper taxi route.
   e. Demonstrates proficiency in maintaining correct and positive control of the airplane's direction and speed considering existing conditions, and uses differential power, when necessary.
   f. Positions flight controls properly considering wind.
   g. Maintains awareness of the location and movement of all other aircraft and vehicles along the taxi path and in the traffic pattern.
   h. Applies right-of-way rules and provides adequate spacing.

i. Avoids creating hazards to persons or property.

2. Action. The examiner will:

a. Ask the applicant to explain taxi procedures.
b. Observe the applicant's taxi procedures, and determine that the applicant's performance meets the objective.

E. TASK: PRETAKE-OFF CHECK (AMEL)

PILOT OPERATION - 1

REFERENCES: AC 61-21; Pilot's Operating Handbook and FAA-Approved Airplane Flight Manual.

1. Objective. To determine that the applicant:

a. Exhibits commercial pilot knowledge of the pre-takeoff check by thoroughly explaining the reasons for checking the items and how to detect possible malfunctions.
b. Positions the airplane properly considering the surface, possible hazards, and wind.
c. Divides attention inside and outside of the cockpit.
d. Ensures that the engine temperatures and pressures are suitable for run-up and takeoff, and avoids any tendency to overheat the engine.
e. Performs a critical and systematic check by following the checklist.
f. Adjusts each control or switch as prescribed by the checklist.
g. Ensures that the airplane is in safe operating condition emphasizing -

(1) flight controls and instruments.
(2) instruments in normal operating range.
(3) engine and propeller operation.
(4) carburetor ice check, if applicable.
(5) fuel valves positioned properly.
(6) seats adjusted and locked for all occupants.
(7) safety belts and shoulder harnesses fastened and adjusted for all occupants.
(8) doors and windows secured.

h. Recognizes indications of any discrepancy and accurately judges whether the airplane is safe for flight or requires maintenance.
i. Reviews the critical takeoff performance airspeeds and expected takeoff distances.
j. Describes takeoff emergency procedures with emphasis on -

(1) engine inoperative cockpit procedures.
(2) engine inoperative airspeeds.
(3) engine inoperative route to follow considering obstructions and wind conditions.

k. Obtains and interprets takeoff and departure clearances.

2. Action. The examiner will:

a. Ask the applicant to explain the reasons for checking the items on the pre-takeoff check and how to detect possible malfunctions.
b. Observe the pre-takeoff check, and determine that the applicant's performance meets the objective.
c. Place emphasis on the applicant's judgment in determining that the airplane is safe for flight.

IV. AREA OF OPERATION: AIRPORT AND TRAFFIC PATTERN OPERATIONS

NOTE: Evaluation in this area is NOT required for additional class-rating applicants.

A. TASK: RADIO COMMUNICATIONS AND ATC LIGHT SIGNALS (AMEL)

PILOT OPERATION - 3

REFERENCE: AC 61-21, AC 61-23; AIM.

1. Objective. To determine that the applicant exhibits commercial pilot competency in radio communications and ATC light signal interpretation including -

a. Selecting the appropriate frequencies for the facilities to be used.
b. Transmitting requests and reports correctly using the recommended standard phraseology.
c. Receiving, acknowledging, and complying with radio communications.
d. Using prescribed procedures following radio communications failure.

2. Action. The examiner will:

a. Observe the applicant's performance in radio communications and ATC light signal interpretation, and determine that the applicant's performance meets the objective.
b. Determine the applicant's effective use of radio communications.

B. TASK: TRAFFIC PATTERN OPERATIONS (AMEL)

PILOT OPERATION - 3

REFERENCES: AC 61-21, AC 61-23; AIM.

1. Objective. To determine that the applicant exhibits commercial pilot competency during traffic pattern operation at controlled and uncontrolled airports including:

a. Collision and wind-shear avoidance procedures.
b. Following the established traffic pattern procedures correctly and consistently adhering to instructions or rules.
c. Correcting for wind-effect to follow the appropriate ground track.
d. Maintaining adequate spacing from other traffic.
e. Maintaining the traffic pattern altitude, ±100 feet.
f. Maintaining the special airspeed, ±10 knots.
g. Completing the pre-landing cockpit checklist.
h. Maintaining orientation with the runway in use.

2. Action. The examiner will observe the applicant's performance in traffic pattern operations, and determine that the applicant's performance meets the objective.

C. TASK: AIRPORT AND RUNWAY MARKING AND LIGHTING (AMEL)

PILOT OPERATION - 3

REFERENCES: AC 61-21; AIM.

1. Objective. To determine that the applicant exhibits commercial pilot competency by:

a. Identifying, interpreting, and conforming to airport, runway, and taxiway marking aids.
b. Identifying, interpreting, and conforming to airport lighting aids.

2. Action. The examiner will observe the applicant's performance in conforming to airport and runway marking and lighting aids, and determine that the applicant's performance meets the objective.

V. AREA OF OPERATION: TAKEOFFS AND CLIMBS

A. TASK: NORMAL AND CROSSWIND TAKEOFFS AND CLIMBS (AMEL)

PILOT OPERATION - 3

REFERENCE: AC 61-21.

1. Objective. To determine that the applicant:

a. Exhibits commercial pilot knowledge by explaining the elements of normal and crosswind takeoffs and climbs including airspeeds, configurations, and emergency procedures.
b. Adjusts the mixture control as recommended for the existing conditions. (The term "recommended" refers to the manufacturer's recommendation. If the manufacturer's recommendation is not available, the description contained in AC 61-21 will be used.)

c. Notes any obstructions or other hazards in the takeoff path and reviews takeoff performance.
d. Verifies wind condition.
e. Aligns the airplane on the runway centerline.
f. Applies aileron deflection in the proper direction, as necessary.
g. Advances the throttles smoothly and positively to maximum allowable power.
h. Checks engine instruments.
i. Maintains positive directional control on the runway centerline.
j. Adjusts aileron deflection during acceleration, as necessary.
k. Rotates at the airspeed to attain lift-off at V_{MC} +5, V_{SSE}, or the recommended lift-off airspeed and establishes wind-effect correction, as necessary.
l. Establishes the single-engine, best rate-of-climb pitch attitude and accelerates to V_Y.
m. Establishes the all-engine best rate-of-climb pitch attitude when reaching V_Y and maintains V_Y, or V_Y +10 knots to avoid high pitch angles.
n. Retracts the wing flaps as recommended or at a safe altitude.
o. Retracts the landing gear after a positive rate of climb has been established and a safe landing cannot be accomplished on the remaining runway, or as recommended.
p. Climbs at V_Y to 400 feet or to a safe maneuvering altitude.
q. Maintains takeoff power to a safe maneuvering altitude and sets desired power.
r. Uses noise abatement procedures, as required.
s. Establishes and maintains a cruise climb airspeed, ±5 knots.
t. Maintains a straight track over the extended runway centerline until a turn is required.
u. Completes the after-takeoff checklist.

2. Action. The examiner will:

a. Ask the applicant to explain the elements of normal and crosswind takeoffs and climbs including airspeeds, configurations, and emergency procedures.
b. Ask the applicant to perform normal and crosswind takeoffs and climbs, and determine that the applicant's performance meets the objective.

B. TASK: MAXIMUM PERFORMANCE TAKEOFF AND CLIMB (AMEL)

PILOT OPERATION - 4

REFERENCE: AC 61-21.

1. Objective. To determine that the applicant:

a. Exhibits commercial pilot knowledge by explaining the elements of a short-field takeoff and climb profile including the significance of appropriate airspeeds, configurations, emergency procedures, and expected performance for existing operating conditions.
b. Selects the recommended wing flap setting.
c. Adjusts the mixture controls as recommended for the existing conditions.
d. Reviews takeoff performance capabilities considering obstructions and conditions affecting the airplane's performance.
e. Positions the airplane for maximum runway availability and aligns it with the runway centerline.
f. Advances throttles smoothly and positively to maximum allowable power.
g. Checks engine instruments.
h. Adjusts the pitch attitude to attain maximum rate of acceleration.
i. Maintains positive directional control on the runway centerline.
j. Rotates at the airspeed to attain lift-off at V_{MC} +5 knots, V_X, VSSE, or at the recommended airspeed, whichever is greater.
k. Climbs at V_X, VSSE, or the recommended airspeed, whichever is greater until obstacle is cleared, or to at least 50 feet above the surface, then accelerates to V_Y and maintains V_Y, or V_Y +10 knots to avoid high pitch angles.
l. Retracts the wing flaps as recommended or at a safe altitude.
m. Retracts the landing gear after a positive rate of climb has been established and a safe landing cannot be made on the remaining runway or as recommended.
n. Climbs at V_Y to 400 feet AGL or to a safe maneuvering altitude.
o. Maintains takeoff power to a safe maneuvering altitude and sets desired power.
p. Uses noise abatement procedures as required.
q. Establishes and maintains a cruise climb airspeed, ±5 knots.
r. Maintains a straight track over the extended runway centerline until a turn is required.
s. Completes the after-takeoff checklist.

2. Action. The examiner will:

a. Ask the applicant to explain the elements of a short-field takeoff and climb including the significance of appropriate airspeeds and configurations, emergency procedures, and the expected performance.
b. Ask the applicant to perform a short-field takeoff and climb, and determine that the applicant's performance meets the objective.

VI. AREA OF OPERATION: INSTRUMENT FLIGHT

NOTE: If an applicant holds a private or commercial pilot certificate with airplane single-engine land and instrument ratings and seeks to add an airplane multiengine land rating, the applicant is required to demonstrate competency in all TASKS of Area of Operation VI.

If the applicant elects not to demonstrate competency in instrument flight, the applicant's multiengine privileges will be limited to VFR only. To remove this restriction, the pilot must demonstrate competency in all TASKS of Area of Operation VI.

If the applicant elects to demonstrate competency in the TASKS of Area of Operation VI, then fails one or more of those TASKS, the applicant will have failed the practical test. After the test is initiated, the applicant will not be permitted to revert to the "VFR only" option.

A. TASK: ENGINE FAILURE DURING STRAIGHT-AND-LEVEL FLIGHT AND TURNS (AMEL)

PILOT OPERATION - 6

REFERENCES: AC 61-21, AC 61-27.

1. Objective. To determine that the applicant:

a. Exhibits commercial pilot knowledge by explaining the reasons for the procedures used if engine failure occurs during straight-and-level flight and during turns while on instruments.
b. Recognizes engine failure promptly during straight-and-level flight and during standard-rate turns.
c. Sets the engine controls, reduces drag, and identifies and verifies the inoperative engine.
d. Establishes the best engine inoperative airspeed and trims the airplane.
e. Verifies the accomplishment of prescribed checklist procedures for securing the inoperative engine.
f. Establishes and maintains a bank toward the operating engine, as necessary, for best performance in straight-and-level flight.
g. Maintains a bank angle, as necessary, for best performance in a turn of approximately standard rate.
h. Attempts to determine the reason for the engine malfunction.
i. Maintains an altitude or a minimum sink rate sufficient to continue flight considering -

(1) density altitude.
(2) service ceiling.
(3) gross weight.
(4) elevation of terrain and obstructions.

j. Monitors the operating engine and makes necessary adjustments.
k. Maintains the specified altitude ±100 feet, if within the airplane's

capability, the specified airspeed ±10 knots, and the specified heading ±10°, if in straight flight.

l. Recognizes the airplane's performance capability and decides an appropriate action to ensure a safe landing.

m. Avoids imminent loss of control or attempted flight contrary to the single-engine operating limitations of the airplane.

2. Action. The examiner will:

a. Ask the applicant to explain the reasons for the procedures used if engine failure occurs during straight-and-level flight and turns.

b. Simulate an engine failure during straight-and-level flight and during standard-rate turns, and determine that the applicant's performance meets the objective.

c. Place emphasis on the applicant's correct performance of emergency procedures on instruments including maintaining the turn if engine failure occurs during this maneuver.

B. TASK: INSTRUMENT APPROACH—ALL ENGINES OPERATING (AMEL)

PILOT OPERATION - 6

REFERENCES: AC 61-21, AC 61-27.

1. Objective. To determine that the applicant:

a. Exhibits commercial pilot knowledge of cockpit management used for a published instrument approach.

b. Requests and receives an actual or a simulated clearance for an instrument approach.

c. Follows instructions and instrument approach procedures correctly.

d. Determines the appropriate rate of descent considering wind and the designated missed approach point.

e. Descends on course so as to arrive at the DH or MDA, whichever is appropriate, in a position from which a normal landing can be made straight-in or circling.

f. Maintains the specified airspeed, ±10 knots.

g. Avoids full-scale deflection on the CDI or glide slope indicators, descent below minimums, or exceeding the radius of turn as dictated by the visibility minimums for the aircraft approach category, while circling.

h. Executes a missed approach at the designated missed approach point and follows appropriate checklist items for airplane cleanup.

i. Communicates properly with ATC.

2. Action. The examiner will:

   a. Ask the applicant to explain the multiengine procedures used for an instrument approach with all engines operating.
   b. Ask the applicant to perform an instrument approach and missed approach with all engines operating, and determine that the applicant's performance meets the objective.

C. TASK. INSTRUMENT APPROACH - ONE ENGINE INOPERATIVE (AMEL)

PILOT OPERATION - 6

REFERENCES: AC 61-21, AC 61-27.

1. Objective. To determine that the applicant:

   a. Exhibits commercial pilot knowledge by explaining the multiengine procedures used during a published instrument approach with one engine inoperative.
   b. Requests and receives an actual or simulated clearance for a published instrument approach.
   c. Recognizes engine failure promptly.
   d. Sets the engine controls, reduces drag, and identifies and verifies the inoperative engine.
   e. Establishes the best engine inoperative airspeed and trims the airplane.
   f. Verifies the accomplishment of the prescribed checklist procedures for securing the inoperative engine.
   g. Establishes and maintains a bank toward the operating engine, as necessary, for best performance.
   h. Attempts to determine the reason for the engine malfunction.
   i. Requests and receives an actual or simulated clearance for a published instrument approach with one engine inoperative.
   j. Follows instructions and instrument approach procedures.
   k. Recites the missed approach procedure and decides on the point at which the approach will continue or discontinue considering the performance capability of the airplane.
   l. Descends on course so as to arrive at the DH or MDA, whichever is appropriate, in a position from which a normal landing can be made straight-in or circling.
   m. Maintains the specified airspeed, ±10 knots.
   n. Avoids full-scale deflection on the CDI or glide slope indicators, descent below minimums, or exceeding the radius of turn as dictated by the visibility minimums for the aircraft approach category, while circling.
   o. Communicates properly with ATC.
   p. Completes a safe landing.

2. Action. The examiner will:

   a. Ask the applicant to explain multiengine procedures used for an instrument approach with one engine inoperative.
   b. Simulate an engine failure and determine that the applicant's performance meets the objective.

VII. AREA OF OPERATION: FLIGHT AT CRITICALLY SLOW AIR SPEEDS

NOTE: The high pitch angles required to induce a stall while using high power settings make controllable flight difficult. For this reason, full stalls have been deleted from this practical test. Imminent stalls only will be evaluated.

Imminent stalls will not be performed with one engine at reduced power or inoperative and the other engine(s) developing effective power.

Examiners and instructors should be alert to the possible development of high sink rates when performing stalls in multiengine airplanes with high wing loadings; therefore, a maximum loss of 50 feet during stall entries has been incorporated in these TASKS.

A. TASK. IMMINENT STALLS, GEAR UP AND FLAPS UP (AMEL)

PILOT OPERATION - 2

REFERENCES: AC 61-21.

1. Objective. To determine that the applicant:

   a. Exhibits commercial pilot knowledge by explaining the aerodynamic factors associated with stalls, gear up and flaps up including changes in stall speed in various configurations, power settings, pitch attitudes, weights, and bank angles, and the procedure for recovery.
   b. Selects an entry altitude that will allow recoveries to be completed no lower than 3,000 feet AGL.
   c. Stabilizes the airplane at approach airspeed in level flight with a gear-up, flaps-up configuration and appropriate power setting.
   d. Establishes straight-and-level flight and level 20° bank turns (±10°) and adjusts pitch attitude and power as necessary to induce an imminent stall while maintaining altitude (+150 feet –50 feet).
   e. Recognizes imminent stalls at the first indication of buffeting or decay of control effectiveness and recovers with proper power and control application.
   f. Returns to airspeed and configuration as specified by the examiner.
   g. Avoids full stall, excessive pitch change, excessive altitude loss, or flight below 3,000 feet AGL.

2. Action. The examiner will:

   a. Ask the applicant to explain the aerodynamic factors associated with stalls, gear up and flaps up.
   b. Ask the applicant to perform imminent stalls with gear up and flaps up in straight flight and turning flight, and determine that the applicant's performance meets the objective.

B. TASK. IMMINENT STALLS, GEAR DOWN AND APPROACH FLAPS (AMEL)

PILOT OPERATION - 2

REFERENCES: AC 61-21.

1. Objective. To determine that the applicant:

   a. Exhibits commercial pilot knowledge by explaining the aerodynamic factors associated with stalls, gear down and approach flaps, including changes in stall speed in various configurations, power settings, pitch attitudes, weights, and bank angles, and the procedure for recovery.
   b. Selects an entry altitude that will allow recoveries to be completed no lower than 3,000 feet AGL.
   c. Stabilizes the airplane at approach airspeed in level flight with gear down and approach flap configuration and appropriate power setting.
   d. Establishes straight-and-level flight and level 20° bank turns, ±10°, and adjusts pitch attitude and power as necessary to induce an imminent stall while maintaining altitude, +150 feet –50 feet.
   e. Recognizes imminent stalls at the first indication of buffeting or decay of control effectiveness and recovers with proper power and control application.
   f. Returns to airspeed and configuration as specified by the examiner.
   g. Avoids full stalls, excessive pitch change, excessive altitude loss, or flight below 3,000 feet AGL.

2. Action. The examiner will:

   a. Ask the applicant to explain aerodynamic factors associated with stalls, gear down and approach flaps.
   b. Ask the applicant to perform imminent stalls with gear down and approach flaps in straight flight and turning flight, and determine that the applicant's performance meets the objective.

C. TASK. IMMINENT STALLS, GEAR DOWN AND FULL FLAPS (AMEL)

PILOT OPERATION - 2

REFERENCES: AC 61-21.

1. Objective. To determine that the applicant:

a. Exhibits commercial pilot knowledge by explaining the aerodynamic factors associated with stalls, gear down and full flaps, including changes in stall speed in various configurations, power settings, pitch attitudes, weights, and bank angles, and the procedure for recovery.
b. Selects an entry altitude that will allow recoveries to be completed no lower than 3,000 feet AGL.
c. Stabilize the airplane at approach airspeed in level flight with a gear down and full flaps configuration and appropriate power setting.
d. Establishes straight-and-level flight and level 20° bank turns, ±10°, and adjusts pitch attitude and power as necessary to induce an imminent stall while maintaining altitude, +150 feet –50 feet.
e. Recognizes imminent stalls at the first indication of buffeting or decay of control effectiveness and recovers with proper power and control application.
f. Returns to airspeed and configuration as specified by the examiner.
g. Avoids full stalls, excessive pitch change, excessive altitude loss, or flight below 3,000 feet AGL.

2. Action. The examiner will:

a. Ask the applicant to explain the aerodynamic factors associated with stalls, gear down and full flaps.
b. Ask the applicant to perform imminent stalls with gear down and full flaps in straight flight and turning flight, and determine that the applicant's performance meets the objective.

D. TASK. MANEUVERING DURING SLOW FLIGHT (AMEL)

PILOT OPERATION - 2

REFERENCES: AC 61-21.

1. Objective. To determine that the applicant:

a. Exhibits commercial pilot knowledge by explaining the flight characteristics and controllability associated with maneuvering during slow flight.
b. Selects an entry altitude that will allow the maneuver to be performed no lower than 3,000 feet AGL.
c. Establishes and maintains slow flight, specified gear position, various flap settings and angle of bank, during straight-and-level flight and level turns.
d. Maintains the specified altitude, ±100 feet.
e. Maintains the specified heading during straight flight, ±10°.
f. Maintains the specified bank angle, ±10°, during turning flight.
g. Maintains an airspeed of 10 knots (±5 knots) above stall speed or V_{MC}, whichever is greater.

2. Action. The examiner will:

   a. Ask the applicant to explain the flight characteristics and controllability involved in slow flight.
   b. Ask the applicant to perform slow flight, specifying the configuration and flight maneuver, and determine that the performance meets the objective.

VIII. AREA OF OPERATION: MAXIMUM PERFORMANCE MANEUVERS

A. TASK: STEEP POWER TURNS (AMEL)

PILOT OPERATION - 4

REFERENCE: AC 61-21.

1. Objective. To determine that the applicant:

   a. Exhibits commercial pilot knowledge by explaining the performance factors associated with steep power turns including load factor and angle-of-bank limitations, effect on stall speed, power required, and overbanking tendency.
   b. Selects an altitude that will allow the maneuver to be performed no lower than 3,000 feet AGL.
   c. Establishes the recommended entry airspeed.
   d. Enters a 360° turn maintaining a bank angle of at least 45°, +10°, –0°, in smooth, stabilized, coordinated flight.
   e. Recognizes the need to apply smooth, coordinated control to maintain the specified altitude, ±100 feet, and the specified airspeed, ±10 knots.
   f. Divides attention between airplane control and orientation.
   g. After completing a 360° turn, reverses direction of turn at the entry heading, ±10°, and performs a 360° turn, then rolls out at the entry heading, ±10°.
   h. Avoids any indication of an approaching stall or tendency to exceed the structural limits of the airplane during the turns.

2. Action. The examiner will:

   a. Ask the applicant to explain the performance factors associated with steep power turns.
   b. Ask the applicant to perform steep power turns and specify the entry heading and amount of turn, and determine that the applicant's performance meets the objective.

IX. AREA OF OPERATION: FLIGHT BY REFERENCE TO GROUND OBJECTS

A. TASK: EIGHTS AROUND PYLONS (AMEL)

PILOT OPERATION - 4

REFERENCE: AC 61-21.

1. Objective. To determine that the applicant:

   a. Exhibits commercial pilot knowledge by explaining the procedures associated with eights around pylons and wind-effect correction throughout the maneuver.
   b. Selects suitable ground reference points that will permit a brief period of straight-and level flight between the pylons.
   c. Enters the maneuver in the proper direction and altitude (600 to 1,000 feet AGL) and a bank angle of approximately 30° to 40° at the steepest point.
   d. Divides attention between coordinated airplane control and ground track.
   e. Applies the necessary wind-effect corrections to track a constant distance from each pylon.
   f. Maintains the specified altitude, ±100 feet, and airspeed, ±10 knots.

2. Action. The examiner will:

   a. Ask the applicant to explain the procedures associated with eights around pylons and necessary wind-effect corrections.
   b. Ask the applicant to perform eights around pylons, and determine that the applicant's performance meets the objective.
   c. Place emphasis on the applicant's planning, coordination, airplane control, wind-effect correction, and division of attention.

X. AREA OF OPERATION: EMERGENCY OPERATIONS

A. TASK: SYSTEMS AND EQUIPMENT MALFUNCTIONS (AMEL)

PILOT OPERATION - 6

REFERENCES: AC 61-21; Pilot's Operating Handbook and FAA-Approved Airplane Flight Manual.

1. Objective. To determine that the applicant:

   a. Exhibits commercial pilot knowledge by explaining causes, indications, and pilot actions for various systems and equipment malfunctions.
   b. Analyzes the situation and takes appropriate action for simulated emergencies such as -

      (1) partial power loss.
      (2) engine roughness or overheat.
      (3) loss of oil pressure.
      (4) carburetor or induction system icing.
      (5) fuel starvation.
      (6) fire in flight.

(7) electrical system malfunction.
(8) hydraulic system malfunction.
(9) landing gear or wing flap malfunction.
(10) door opening in flight.
(11) trim inoperative.
(12) pressurization system malfunction.
(13) other malfunctions.

2. Action. The examiner will:

a. Ask the applicant to explain causes, indications, and remedial action for various systems and equipment malfunctions.
b. Simulate various equipment malfunctions, and determine that the applicant's performance meets the objective.
c. Place emphasis on the applicant's ability to analyze the situation and take action appropriate to the simulated emergency.

B. TASK: MANEUVERING WITH ONE ENGINE INOPERATIVE (AMEL)

PILOT OPERATION - 6

REFERENCES: AC 61-21; Pilot's Operating Handbook and FAA-Approved Airplane Flight Manual.

NOTE: The feathering of one propeller should be demonstrated in any multiengine airplane equipped with propellers which can be safely feathered and unfeathered in flight. Feathering for pilot flight test purposes should be performed only under such conditions and at such altitudes (no lower than 3,000 feet above the surface) and positions where safe landings on established airports can be readily accomplished should difficulty be encountered in unfeathering. At altitudes lower than 3,000 feet above the surface, simulated engine failure will be performed by reducing the selected engine to zero thrust.

A propeller that cannot be unfeathered during the practical test should be treated as an emergency.

1. Objective. To determine that the applicant:

a. Exhibits commercial pilot knowledge by explaining the flight characteristics and controllability associated with maneuvering with one engine inoperative.
b. Sets the engine controls, reduces drag, identifies and verifies the inoperative engine after simulated engine failure.
c. Attains the best engine inoperative airspeed and trims the airplane.
d. Maintains control of the airplane.
e. Attempts to determine the reason for the engine malfunction.

f. Follows the prescribed checklist to verify procedures for securing the inoperative engine.
g. Establishes a bank toward the operating engine, as necessary, for best performance.
h. Turns toward the nearest suitable airport.
i. Monitors the operating engine and makes necessary adjustments.
j. Demonstrates coordinated flight with one engine inoperative (propeller feathered, if possible) including -

(1) straight-and-level flight.
(2) turns in both directions.
(3) descents to assigned altitudes.
(4) climb to assigned altitudes, if airplane is capable of climbs under existing conditions.

k. Maintains the specified altitude, ±100 feet, when a constant altitude is specified, and levels off from climbs and descents, at specified altitudes, ±100 feet.
l. Maintains the specified heading during straight flight, ±10°.
m. Maintains the specified bank angle, ±10°, during turns.
n. Divides attention between coordinated control, flight path, and orientation.
o. Demonstrates engine restart in accordance with prescribed procedures.

2. Action. The examiner will:

a. Ask the applicant to explain the flight characteristics and controllability involved in flight with one engine inoperative.
b. Simulate engine failure and observe the procedures, and determine that the applicant's performance meets the objective.

C. TASK: ENGINE INOPERATIVE LOSS OF DIRECTIONAL CONTROL DEMONSTRATION (AMEL)

PILOT OPERATION - 6

REFERENCES: AC 61-21; Pilot's Operating Handbook and FAA-Approved Airplane Flight Manual.

NOTE: There is a density altitude above which the stalling speed is higher than the engine inoperative minimum control speed. When this density altitude exists close to the ground because of high elevations and/or high temperatures, an effective flight demonstration of loss of directional control may be hazardous and should not be attempted. If it is determined prior to flight that the stall speed is higher than V_{MC} and this flight demonstration is impracticable, the significance of the engine inoperative minimum control speed should be emphasized through oral questioning, including the results of attempting engine inoperative flight below this speed, the recognition of loss of directional control, and proper recovery techniques.

To conserve altitude during the engine inoperative loss of directional control demonstration, recovery should be made by reducing angle of attack and resuming controlled flight. If a situation exists where reduction of power on the operating engine is necessary to maintain airplane control, the decision to reduce power must be made by the pilot to avoid uncontrolled flight. Emphasis should be placed on the safe conservation of altitude.

Recoveries should never be made by increasing power on the simulated failed engine.

The practice of entering this maneuver by increasing pitch attitude to a high point with both engines operating and then reducing power on the critical engine should be avoided because the airplane may become uncontrollable when the power on the critical engine is reduced.

1. Objective. To determine that the applicant:

   a. Exhibits commercial pilot knowledge by explaining the causes of loss of directional control at airspeeds less than V_{MO} minimum engine inoperative control speed), the factors affecting V_{MC}, and the safe recovery procedures.
   b. Selects an entry altitude that will allow recoveries to be completed no lower than 3,000 feet AGL.
   c. Establishes the airplane configuration with -

      (1) propeller set to high RPM.
      (2) landing gear retracted.
      (3) flaps set in takeoff position.
      (4) cowl flaps set in takeoff position.
      (5) engines set to rated takeoff power or as recommended.
      (6) trim set for takeoff.
      (7) power on the critical engine reduced to idle (avoid abrupt power reduction).

   d. Establishes a single-engine climb attitude (inoperative engine propeller windmilling) with the airspeed representative of that following a normal takeoff.
   e. Establishes a bank toward the operating engine, as necessary, for best performance.
   f. Reduces the airspeed slowly with the elevators while applying rudder to maintain directional control until all available rudder is applied.

g. Recognizes the indications of loss of directional control.
h. Recovers promptly by reducing the angle of attack to regain control and, if necessary, adjusts power on operating engine sufficiently to maintain control with minimum loss of altitude.
i. Recovers to the entry heading, ±10°.

2. Action. The examiner will:

a. Ask the applicant to explain the causes of loss of directional control, the factors affecting V_{MC}, and safe recovery procedures.
b. Ask the applicant to demonstrate engine inoperative loss of directional control, and determine that the applicant's performance meets the objective.

D. TASK: DEMONSTRATING THE EFFECTS OF VARIOUS AIRSPEEDS AND CONFIGURATIONS DURING ENGINE INOPERATIVE PERFORMANCE (AMEL)

PILOT OPERATION - 6

REFERENCES: AC 61-21; Pilot's Operating Handbook and FAA-Approved Airplane Flight Manual.

1. Objective. To determine that the applicant:

a. Exhibits commercial pilot knowledge by explaining the effects of various airspeeds and configurations on performance during engine inoperative operation.
b. Selects an entry altitude that will allow recoveries to be completed no lower than 3,000 feet AGL.
c. Establishes V_{YSE} with critical engine at zero thrust.
d. Varies the airspeed from V_{YSE} and demonstrates the effect of the airspeed changes on performance.
e. Maintains V_{YSE} and demonstrates the effect of each of the following on performance -

(1) extension of landing gear.
(2) extension of wing flaps.
(3) extension of both landing gear and wing flaps.
(4) windmilling of propeller on the critical engine.

2. Action. The examiner will:

a. Ask the applicant to explain the effects of various airspeeds and various configurations on performance during engine inoperative operation.
b. Ask the applicant to demonstrate the effects of various airspeeds and various configurations on performance, and determine that the applicant's performance meets the objective.

E. TASK: ENGINE FAILURE ON TAKEOFF BEFORE V_{MC} (AMEL)

PILOT OPERATION - 6

REFERENCES: AC 61-21; Pilot's Operating Handbook and FAA-Approved Airplane Flight Manual.

1. Objective. To determine that the applicant:
   a. Exhibits commercial pilot knowledge by explaining the reasons for the procedures used for engine failure during takeoff before V_{MC} including related safety factors.
   b. Aligns the airplane on the runway centerline.
   c. Advances the throttles smoothly to maximum allowable power.
   d. Checks engine instruments.
   e. Maintains directional control on the runway centerline.
   f. Closes throttles smoothly and promptly when engine failure occurs.
   g. Maintains directional control and applies braking, as necessary.

2. Action. The examiner will:
   a. Ask the applicant to explain the reasons for the procedures used for engine failure during takeoff before V_{MC} including related safety factors.
   b. Ask the applicant to perform a takeoff and will reduce power on an engine before reaching 50 percent V_{MC}, and determine that the applicant's performance meets the objective.

F. TASK: ENGINE FAILURE AFTER LIFT-OFF (AMEL)

PILOT OPERATION - 6

REFERENCES: AC 61-21; Pilot's Operating Handbook and FAA-Approved Airplane Flight Manual.

1. Objective. To determine that the applicant:
   a. Exhibits commercial pilot knowledge by explaining the reasons for the procedures used if engine failure occurs after lift-off including related safety factors.
   b. Recognizes engine failure promptly.
   c. Sets the engine controls, reduces drag, and identifies and verifies the inoperative engine after simulated engine failure.
   d. Establishes V_{YSE} if there are no obstructions; if obstructions are present, establishes V_{XSE} or V_{MC} +5, whichever is greater, until obstructions are cleared, then V_{YSE} and trims the airplane.
   e. Maintains positive control of the airplane.
   f. Follows the prescribed checklist to verify the accomplishment of procedures for securing the inoperative engine.

g. Establishes a bank toward the operating engine as required for best performance.
h. Recognizes the airplane's performance capability; if climb or level flight is impossible, maintains V_{YSE} and initiates an approach to the most suit able landing area.
i. Attempts to determine the reason for the engine malfunction.
j. Monitors the operating engine and makes necessary adjustments.
k. Maintains the specified heading, ±10°, and the specified airspeed, ±5 knots.
l. Divides attention between coordinated airplane control, flight path, and orientation.
m. Contacts the appropriate facility for assistance, if necessary.

2. Action. The examiner will:

a. Ask the applicant to explain the reasons for the procedures used for engine failure after lift-off including related safety factors.
b. Simulate engine failure after lift-off, considering all safety factors, by retarding the throttle to zero thrust, and determine that the applicant's performance meets the objective.

G. TASK: ENGINE FAILURE EN ROUTE (AMEL)

PILOT OPERATION - 6

REFERENCES: AC 61-21; Pilot's Operating Handbook and FAA-Approved Airplane Flight Manual.

1. Objective. To determine that the applicant:

a. Exhibits commercial pilot knowledge by explaining the techniques and procedures used if engine failure occurs while en route.
b. Sets the engine controls, reduces drag, and identifies and verifies the inoperative engine after simulated engine failure.
c. Attains the best engine inoperative airspeed and trims the airplane.
d. Maintains control of the airplane.
e. Attempts to determine the reason for the engine malfunction.
f. Follows the prescribed checklist to verify the accomplishment of procedures for securing the inoperative engine.
g. Establishes a bank toward the operating engine, as necessary, for best performance.
h. Turns toward nearest suitable airport.
i. Maintains an altitude or a minimum sink rate sufficient to continue flight considering -

(1) density altitude.

(2) service ceiling.
(3) gross weight.
(4) elevation of terrain and obstructions.

j. Monitors the operating engine and makes necessary adjustments.
k. Maintains the specified altitude, ±100 feet, if within the airplane's capability, the specified heading, ±10°, and the specified airspeed, ±5 knots.
l. Divides attention between coordinated airplane control, flight path, and orientation.
m. Contacts appropriate facility for assistance, if necessary.

2. Action. The examiner will:

a. Ask the applicant to explain the techniques and procedures used for engine failure en route.
b. Simulate an engine failure while en route and observe the applicant's ability to follow prescribed procedures, and determine that the applicant's performance meets the objective.

H. TASK: APPROACH AND LANDING WITH AN INOPERATIVE ENGINE (AMEL)

PILOT OPERATION - 6

REFERENCES: AC 61-21; Pilot's Operating Handbook and FAA-Approved Airplane Flight Manual.

1. Objective. To determine that the applicant:

a. Exhibits commercial pilot knowledge by explaining the procedure used during an approach and landing with an inoperative engine.
b. Sets the engine controls, reduces drag, and identifies and verifies inoperative engine after simulated engine failure.
c. Establishes the recommended airspeed and trims the airplane.
d. Follows the prescribed checklist to verify procedures for securing the inoperative engine and completes pre-landing checklist.
e. Establishes a bank toward the operating engine as required for best performance.
f. Maintains proper track on final approach.
g. Establishes the approach and landing configuration and power.
h. Maintains a stabilized descent angle and the recommended final approach airspeed (not less than V_{YSE}) until landing is assured.
i. Touches down smoothly beyond and within 500 feet of a specified point, with no drift and the longitudinal axis aligned with the runway centerline.
j. Maintains positive directional control during after-landing roll.

2. Action. The examiner will:

a. Ask the applicant to explain the procedures used during an approach and landing with an inoperative engine.
b. Simulate an engine failure by setting one engine at zero thrust, and determine that the applicant's performance meets the objective.

XI. AREA OF OPERATION: APPROACHES AND LANDINGS

A. TASK: NORMAL AND CROSSWIND APPROACHES AND LANDINGS (AMEL)

PILOT OPERATION - 3

REFERENCES: AC 61-21; Pilot's Operating Handbook and FAA-Approved Airplane Flight Manual.

1. Objective. To determine that the applicant:

a. Exhibits commercial pilot knowledge by explaining the elements of normal and crosswind approaches and landings including airspeeds, configurations, performance, and related safety factors.
b. Establishes the approach and landing configuration and adjusts the power controls, as required.
c. Maintains a stabilized descent angle and the recommended approach airspeed, with gust factor applied, ±5 knots.
d. Notes any obstructions or other hazards in the approach path and landing area, and considers landing performance capability.
e. Verifies wind condition and makes positive correction for crosswind.
f. Maintains a precise ground track on final approach.
g. Recognizes and promptly corrects deviations during approach and landing.
h. Makes smooth, timely, and precise control application during the transition from approach to landing roundout (flare).
i. Touches down smoothly at approximate stalling speed, beyond and within 200 feet of a specified point, with no drift and the airplane's longitudinal axis aligned with the runway centerline.
j. Maintains positive directional control and crosswind correction during the after-landing roll.
k. Completes the after-landing checklist in a timely manner.

2. Action. The examiner will:

a. Ask the applicant to explain the elements of a normal and crosswind approach and landing, including airspeeds, configurations, landing performance, and related safety factors.
b. Ask the applicant to perform a normal and a crosswind approach and landing in various flap configurations, and determine that the applicant's performance meets the objective.

B. TASK: GO-AROUND FROM REJECTED (BALKED) LANDING (AMEL)

PILOT OPERATION - 3

REFERENCES: AC 61-21; Pilot's Operating Handbook and FAA-Approved Airplane Flight Manual.

1. Objective. To determine that the applicant:

   a. Exhibits commercial pilot knowledge by explaining the elements of a go-around procedure, including the recognition of the need to go around, the importance of making a timely decision, the use of recommended airspeeds, the drag effect of wing flaps and landing gear, and the importance of properly coping with undesirable pitch and yaw tendencies.
   b. Makes a timely decision to go around from a rejected landing.
   c. Applies takeoff power and establishes the precise pitch attitude required to attain the recommended airspeed.
   d. Retracts the wing flaps, as recommended, or at a safe altitude, and establishes V_Y.
   e. Retracts the landing gear after a positive rate of climb has been established.
   f. Trims the airplane and climbs at V_Y, ±5 knots, and maintains the proper ground track in the traffic pattern.

2. Action. The examiner will:

   a. Ask the applicant to explain the elements of a go-around procedure including the recognition of the need to go around, the importance of making a timely decision, the use of the recommended airspeeds, the drag effect of wing flaps and landing gear, and the importance of coping with undesirable pitch and yaw tendencies.
   b. Establish a situation in which a go-around from a rejected landing would be required, and determine that the applicant's performance meets the objective.
   c. Place emphasis on the applicant's judgment, prompt action, and ability to maintain positive control of the airplane during the go-around.

C. TASK: MAXIMUM PERFORMANCE APPROACH AND LANDING (AMEL)

PILOT OPERATION - 4

REFERENCES: AC 61-21; Pilot's Operating Handbook and FAA-Approved Airplane Flight Manual.

1. Objective. To determine that the applicant:

   a. Exhibits commercial pilot knowledge by explaining the elements of a

short-field approach and landing, including airspeeds, configurations, and related safety factors.
- b. Considers obstructions, landing surface, and wind conditions.
- c. Selects a suitable touchdown point.
- d. Establishes the recommended short-field approach and landing configuration and adjusts power and pitch, as required.
- e. Maintains a stabilized descent angle, precise control of the descent rate, and recommended airspeed.
- f. Maintains a precise ground track on final approach.
- g. Recognizes and promptly corrects deviations during approach or landing.
- h. Makes smooth, timely, and precise control application during the transition from approach to landing roundout (flare).
- i. Touches down smoothly beyond and within 100 feet of a specified point, no drift, and with the airplane longitudinal axis aligned with the runway centerline.
- j. Maintains positive directional control during the after-landing roll.
- k. Applies smooth braking, as necessary, to stop in the shortest distance consistent with safety.
- l. Completes the after-landing checklist in a timely manner.

2. Action. The examiner will:
   - a. Ask the applicant to explain the elements of a short-field approach and landing including airspeeds, configurations, and related safety factors.
   - b. Ask the applicant to perform a short-field approach and landing, and determine that the applicant's performance meets the objective.

D. TASK: AFTER-LANDING PROCEDURES (AMEL)

PILOT OPERATION - 3

REFERENCES: AC 61-21; Pilot's Operating Handbook and FAA-Approved Airplane Flight Manual.

1. Objective. To determine that the applicant:
   - a. Exhibits commercial pilot knowledge by explaining the after-landing procedure including taxiing, parking, shutdown, securing, and post-flight inspection.
   - b. Selects and taxies to the designated or suitable parking area considering wind conditions and obstructions.
   - c. Parks the airplane properly.
   - d. Follows the recommended procedure for engine shutdown, cockpit securing, and deplaning passengers.
   - e. Secures the airplane properly.
   - f. Performs a satisfactory post-flight inspection.

2. Action. The examiner will:

   a. Ask the applicant to explain the after-landing procedures including taxiing, parking, shutdown, securing, and post-flight inspection.
   b. Observe the applicant's after-landing procedures, and determine that the applicant's performance meets the objective.

9
The multiengine flight instructor

IF YOUR CAREER GOALS INVOLVE THE AIRLINES OR CORPORATIONS, YOU will need all the multiengine flight time possible as outlined in chapter 10. Time spent in the multiengine airplane giving instruction to others is usually the first opportunity for young pilots to build multiengine experience.

The bad news is that multiengine flight instruction can be very hazardous. The multiengine instructor must really know the airplane and its capabilities. There are more opportunities to get into trouble while sitting in the right seat of a light multiengine airplane than in any other type of instruction.

PRIMARY INSTRUCTION FIRST

You must graduate to multiengine instruction. I would not want a rookie instructor teaching from the start in a multiengine airplane. So many things learned by giving primary instruction prove to be priceless in a multiengine airplane. Instructors learn to divide their attention with primary students. Experienced instructors do a better job of looking out for traffic. Experienced instructors can keep one eye on the student and one eye on the airplane.

After giving a few hundred hours of instruction, your time in the air (and the student's time) is used more efficiently. You become a better teacher because you have explained many ideas to all types of students. Experienced instructors know what does and does not work. All these lessons learned by teaching in a single-engine airplane are necessary for good, safe, multiengine instruction.

You never stop learning and growing. I fully understood instrument flying only when I became an instrument instructor. Likewise, I fully understood multiengine flying only when I became a multiengine instructor.

Giving multiengine flight instruction is a real challenge. You must teach new concepts to students every day. The students who come to get their multiengine ratings probably have never talked about zero side slip, V_{MC}, accelerate-stop distance, crossfeeding an engine, or feathering a propeller. This will all be new to them. It will be up to you to truly understand these topics and relate the art and science of multiengine flight to them.

INSTRUCTOR CHECKRIDE

A multiengine flight instructor certificate requires another checkride that will be similar to the original multiengine rating test, but now you fly from the right seat and you must be able to teach all the multiengine concepts. Start instructing in a single-engine airplane because training for a flight instructor certificate with a multiengine rating would be very expensive without the single-engine experience. Additionally, you could only instruct in multiengine airplanes.

There is no multiengine instructor written test, so expect a vigorous oral exam to begin the practical test. Regulation 91.191 says that to qualify for the multiengine instructor test, you must have logged a minimum of 15 hours in multiengine airplanes. The 6–10 hours or more that you spent working toward your original multiengine rating do not count toward the 15 because you were not yet rated in a multiengine airplane during your training. This means that the 15 hours must all come after passing your first multiengine checkride.

Later, when you are giving multiengine instruction, the FAA requires under regulation 91.195 that you have 5 hours of pilot-in-command time in any multiengine airplane in which you teach.

The flight test will be easier and shorter if you already have a flight instructor certificate with an airplane single-engine rating. The multiengine instructor practical test standard says that if you have been previously tested on a specified area of operation, then you need not be tested again.

The areas eliminated from your multiengine instructor test if you already have a single-engine instructor certificate are:

- Area of operation I—Fundamentals of instructing.
- Area of operation II—Technical subject areas.
- Area of operation III—Preflight preparation.

- Area of operation IV—Preflight lesson on a maneuver to be performed in flight.
- Area of operation VI—Ground and water operations.
- Area of operation VII—Airport operations.
- Area of operation IX—Fundamentals of flight.
- Area of operation XI—Basic instrument maneuvers.
- Area of operation XII—Performance maneuvers.
- Area of operation XIII—Ground reference maneuvers.
- Area of operation XVI—After landing procedures (seaplane).

Quite a bit is cut out, but realize that according to the standards: "At the discretion of the examiner, the applicant's competence in all areas of operation may be evaluated." This means that the examiner might require an applicant to prepare a lesson plan, discuss the laws of learning, or fly an S turn on the multiengine instructor test.

These areas of operation would become the basis for the multiengine instructor test (edited) if you already have a single-engine instructor certificate:

- Area of operation V—Multiengine operations: operation of systems; performance and limitations; flight principles with an engine inoperative; emergency procedures.
- Area of operation VIII—Takeoffs and climbs: normal and crosswind takeoff and climb; short-field takeoff and climb.
- Area of operation X—Stalls and maneuvering during slow flight: power-on stalls; power-off stalls; maneuvering during slow flight.
- Area of operation XIV—Emergency operations: system and equipment malfunctions; maneuvering with one engine inoperative; engine inoperative loss of directional control demonstration (V_{MC} demo); demonstrating the effects of various airspeeds and configurations during engine inoperative performance; engine failure during takeoff before V_{MC}; engine failure after liftoff; approach and landing with an inoperative engine; emergency equipment and survival gear.
- Area of operation XV—Approaches and landings (airplane): normal and crosswind approach and landing; go-around; short-field approach and landing.

METHOD OF OPERATION

After becoming a multiengine instructor, you will need to have a talk with yourself to determine exactly what your limitations will be. I struck a compromise in the previous chapter on multiengine airwork.

One school of thought says that you should teach total realism in multiengine training. You should actually feather an engine on takeoff, and you should make real engine-failed approaches and landings. The other school of thought never actually shut down an engine in flight. Always simulate the engine failure with the throttle, never the mixture control or the fuel selector valve.

The first school is too dangerous and the second school does not offer real multiengine training. I split the difference at 3,000 feet AGL, or higher if recommended by the manufacturer. Above 3,000 feet AGL, I use the mixture control or selector valve to shut down, feather, and then secure engines. Below 3,000 feet AGL, I use the throttle only to simulate an engine failure.

You must develop some personal minimums as a multiengine instructor. There are certain things that you will not do. Remember when you first got your instrument rating? The new rating in your wallet said that you could legally fly in any instrument weather, but you didn't. If you were smart, your first solo IFR experience was an approach with a relatively high ceiling so that you broke out of the clouds well above the decision height. Your confidence grew, and you became comfortable with lower and lower ceilings.

The same is true with multiengine instruction. At first, you will not even simulate an engine failure upon takeoff below 500 feet AGL. You might lower that to 300 feet after learning more about the airplane and yourself. Only you can know your comfort level. Do not get pushed into a situation that you cannot get out of.

Takeoff engine-failure simulations should be accomplished by bringing one throttle back to idle. The student should go through the procedure in response to a failure. If and when the student calls for prop feathering, I will advance the idle throttle to a position that gives zero thrust, which simulates the drag of a feathered prop. I do not place the throttle to zero thrust until the student calls "prop to feather." If he forgets that part of the procedure, he is going to have sore leg muscles.

Always be ready for your student to do something unexpected. Never hesitate to pull back both mixture controls to save a situation. For instance, you are on takeoff roll with a multiengine student. As a test for the student, prior to obtaining V_{MC} you call out, "Fire in the left engine!" The student, rather than pulling back both throttles, retards only the left one.

If you are not ready, you will soon be off the side of the runway and into the grass. Pull back both mixture controls, and keep the nose straight with rudder. Anticipate. Be 110 percent attentive while providing multiengine instruction.

Would you ever pull both mixtures in flight? Yes, if the alternative is a low altitude V_{MC} roll. You have lifted off the runway with a multiengine student on his third lesson, and you simulate right-engine failure by pulling the right throttle back to idle. Your eager student, his mind a blur, quickly pulls the right engine into feather, and locks his elbow holding full power on the left engine.

It happens so fast that verbal instructions cannot prevent it. The airplane is losing speed and altitude. The airplane starts to yaw to the right. "Pull the rip cord" this instant: both mixture controls. Remember that it's always better to pancake in with the wheels down than to crash upside down and out of control. If this hypothetical situation scares you, you are not alone.

Protect yourself

Do not simulate takeoff engine failures until you have some confidence in your student's airspeed control. Do not simulate takeoff engine failures near V_{MC}. Do not simulate takeoff engine failures immediately as the wheels leave the ground.

Perform engine shutdowns at a safe altitude. You must know the specific airplane well to confidently shut down an engine. You need to know the manufacturer's V_{SSE} speed by heart. V_{SSE} is the *minimum safe single-engine* speed used to intentionally render an engine inoperative for training purposes.

Whenever you intend to shut down and feather an engine, have a plan of action already in mind if the engine does not restart. The airplane's mechanical well-being often determines which engine to shut down. You need to know which engine is easier to start. You need to know which starter does not always engage the first time. You need to know any other factor that would affect an engine restart.

At high altitude, you can keep the identity of the failed engine a secret. Down low, when the throttle is used to simulate engine failure, the student can clearly see which engine is pulled back. Up high, where you do complete engine shutdowns, hide the engine failure from the student. If you fail an engine by using the fuel selector valve, move the valve after distracting the student, or cover up both valves with a chart or checklist.

Cover the valves when you are not going to fail an engine; the student will not know for sure when the engine might fail or which engine it will be. If you fail the engine with the mixture control, you can place a chart or checklist into the control pedestal between the right-engine prop control and the left-engine mixture control. The student will know an engine is about to fail, but he won't know which one.

Think ahead

Think about the wind at the home airport before you shut down an engine. Provided that both engines start equally well, which engine should be shut down in the following situation? Runway 18 is being used at home. The wind is from 220° at 15 knots. You are 5 miles from the airport at 5,000 feet AGL, and you have reached the point in the lesson where you want to teach the student to recognize engine failure, feather the prop, and secure the engine.

The left engine is better. Let's assume that a restart fails and several additional attempts also fail. Now you are faced with actually landing with a single engine. Wind at the airport will require a right crosswind technique. This means the centerline-aligned approach will carry the right wing low. If the left engine is dead, you should fly with the right wing low anyway.

To reduce side-slip drag, you will make the approach with the airplane banked into the good engine. Everything will work out if the good engine is on the same side that the wind is coming from. Imagine how awkward and dangerous the approach would be with the left wing down in a right crosswind. Plan ahead so that this never happens.

It is also a good idea to have some engine-out boundaries. Never let yourself get too far from a suitable airport while practicing engine-out procedures. The distance depends on altitude and wind. If you were just at the minimum 3,000 feet AGL and stuck with an engine that would not restart, you could be in trouble if you must fly to the airport against a strong breeze.

Never attempt to fly slowly or execute a stall with one engine inoperative. Never simulate an engine failure when flying slower than V_{SSE}.

NO RIDERS

Think twice about allowing riders on multiengine flight lessons. Another student might benefit from observation, but other factors are involved. I will not do V_{MC} demonstrations with a backseat passenger. Any extra weight in the rear moves the center of gravity, which increases V_{MC}. The bookspeed for V_{MC} is calculated with an aft CG, but a minimal surprise factor is preferred during V_{MC} work; the calculated speed of V_{MC} and the actual speed of V_{MC} can be quite different.

Instructing might be just what you need to gain that expensive and irreplaceable multiengine time. It can be great fun, but approach it very seriously. Do your homework. Know your airplane inside and out. Be ready to react to student mistakes regardless of the student's experience. Always leave yourself an out.

10
Why multiengine?

WHY IS IT IMPORTANT TO GET A MULTIENGINE RATING? YOU MIGHT HAVE access to a multiengine airplane; learning to fly it would represent a new challenge. For most pilots, the step from single-engine to multiengine is not taken "because it is there." Rather, it is taken as a necessary step on a career path.

It should be understood that climbing the aviation career ladder to corporations or airlines cannot be accomplished without a multiengine rating and multiengine flight time. If a piloting career is your goal, you must cultivate what will make you attractive to companies in the future. Your pilot logbook must become an unshakable legal document that clearly explains the course you have taken to reach the goal.

Not long after I passed the commercial pilot checkride, and with barely 300 hours of flight time, I was introduced at a gathering as a "commercial pilot." A man asked me, "So what airline do you fly for?" It was a little embarrassing trying to explain that just because I had a commercial certificate did not mean that I could fly for a commercial airline. There is a large gap between certification and marketability.

A corporation or airline looks at many factors when they hire an individual: personality, references from previous employers, drug tests, knowledge of airplane systems, regulations and instrument procedures, ease with relationships, previous FAA violations, and many others. But the factor that gets you in the door to the interview in the first place is flight time. Building the flight time that will attract an interview leading to professional pilot employment is a skill all its own.

QUANTITY AND QUALITY FLIGHT TIME

Your flight time must consider two factors: quantity and quality. When a pilot first gets the commercial certificate with as little as 200 hours of flight in the logbook, quantity is more important than quality. This pilot must increase flight time or wither on the vine. The amount of time at this point is more important than what kind of time. The pilot must build flight time from 200 to approximately 1,000 hours. During this stretch, an hour of Cessna 150 time is just as valuable as anything else.

Quantity is most important during this period, but quality should not be ignored. For instance, flights to Atlanta, Georgia, are more valuable than flights to Edenton, North Carolina. The actual time in the logbook might be the same for both flights, but the level of pilot skill required and experience gained from the flight is greater in Atlanta.

A flight in actual instrument conditions is more valuable than a flight in VFR. A flight that deals with northern winter weather is more valuable than a flight into southern summer weather. Flight instruction given is more valuable than a sightseeing flight. A night Part 135 cargo flight is more valuable than a dual cross-country with a student pilot. Anything "scheduled" is more valuable than anything unscheduled.

After the pilot puts together around 1,000 hours, the quantity versus quality equation begins to turn around. Now it becomes more important to have meaningful hours rather than plentiful hours, and this is where multiengine time makes all the difference. A pilot with 2,500 single-engine hours is not as valuable as a pilot with 2,000 hours that include 500 hours in a multiengine airplane. The quantity (2,500) is not as good as the quality (2,000/500). Employers expect the quantity, but look for quality.

FLIGHT TIME MINIMUMS

Airline companies publicly issue minimum flight times for pilots that they will consider for employment. There is some variation between companies, but most lump all multiengine time together. A check airman with American Eagle told me, "All multiengine time is counted the same." Flight time in a Piper Seminole is as good as a Beech Baron. In the interview, the larger equipment might make a difference to the individual conducting the interview, but all reciprocating multiengine time is equal. Turbine flight time does carry extra clout, but first-job employers rarely include turbine time minimums in their published numbers.

I teach a college course in air transportation that details how to get a job in the aviation industry. I assign every student an airline company. The airlines chosen are usually the kind considered to be entry level: commuters and regional carriers. The student is required to do extensive research on that airline so that an educated employment decision could be made about that company.

This is not an ordinary college paper. Students must dig past the company's propaganda to get the real story. They must interview company officials, talk to the pilots, study the company's financial position, determine if airplanes are on order, and learn about the hiring process, including minimum flight time requirements. Additionally, the students learn about education and health requirements, understand the management and union relationship, predict the airline's expansion or bankruptcy, and understand the motivation of the airline to make interline agreements with other carriers. Armed with all this information, a better employment decision can be made.

Every airline requires multiengine time. Here is a sample of airline requirements from research in the early 1990s. Some of these airlines went out of business; others have been swallowed up by other airlines. The figures show a multiengine pattern. The first number is the airline's minimum total flight time. The second number is the airline's minimum multiengine flight time.

- Air LA 1,500/250
- Air Midwest 1,500/300
- Alpha Air 1,200/200
- Aspen Airways 2,500/500
- Business Exp 1,500/500
- CC Air/USAir 1,500/200
- Crown Airways 1,500/500
- Eastern/B Harbour 1,200/200
- GP Express 1,500/500
- Henson/USAir 2,000/500
- Jetstream International 1,500/500
- Long Island Air 1,000/150
- Midwest Express 2,000/1,000
- Mohawk 1,500/500
- New England Air 2,000/500
- Precision 1,500/500
- Simmons 1,500/300
- SkyWest Airlines 1,000/100
- Sunshine Air 3,000/1,000
- TransWorld Exp 1,500/500
- Virgin Air 1,500/200

The times shown were given from the companies themselves at the time of the research. These times can and do change from day to day depending on the company needs. The real reason these numbers are published by the companies is to discourage pilots with less time from flooding their personnel office with résumés and phone calls. There are plenty of cases where pilots were hired with less than the stated minimum times, but normally you will need much more than the stated minimum to compete for the job.

Total time requirements and multiengine requirements have a fairly broad range. Is 1,500 total and 500 multiengine the magic mix of quantity and quality? Maybe not. Remember that these numbers are used to "qualify" candidates for employment. Hiring decisions are also made on factors such as personality, work ethic, and luck.

PILOT EMPLOYMENT ECONOMICS

In the late 1980s, pilots were being hired at an alarming rate by airline companies. Two factors came together to fuel the hiring frenzy. The 1978 deregulation act took the federal government out of the airline business. Prior to deregulation, airline companies were subsidized to fly to certain airports. The government decided that many communities needed airline service so that their populations, job markets, and tax bases would grow.

If an airline company lost money servicing a community, the federal government would simply write them a check to cover the loss. This guaranteed the airlines a profit. This also eliminated the need for one airline to compete with another on the basis of ticket price. The airlines were wealthy, stable, and stationary. This kept ticket prices artificially high. Consumer groups wanted competition in the market to drive prices lower.

After deregulation, if an airline company lost money flying to a particular city, it no longer could turn to the government for dollars. Smart business sense says that a company should always maximize profits or at least minimize losses. With this strategy in mind, many cities lost air service because the airline lost money.

Small airline companies with smaller airplanes came to the rescue. Turboprop commuter airplanes were best suited to make a profit on short hops to a larger airport to connect with larger airlines. More than 500 small airlines sprang up to fill the gaps left by the major carriers. These airlines needed pilots, and when companies need pilots, the market is right for pilot hiring.

At the same time that deregulation was being felt, the military changed its strategy in regard to pilots. Airline pilots traditionally trained and built time in the military, then went to work for an airline. The airlines relied on a steady stream of pilots with previous big airplane experience from the military.

The military saw many of its training dollars lost to the civilian airlines and decided that money would be saved in the long run by increasing military pilots' pay and luring them to stay in the service. It worked. In 1985, for the first time, more civilian-trained pilots were hired by major airlines than military-trained pilots.

I spoke to a Navy pilot flying a P-3 Orion and asked him if he was considering the airlines when his Navy obligation was completed. "I would like to leave the Navy so my family would not get transferred so often, but if I leave the Navy now and go to the airlines it will take me at least five years to make what I already make now. I cannot afford the pay cut, so I'm staying in the Navy."

Airlines needed pilots in the late 1980s, and the traditional military source was drying up. The only other source was civilian pilots, but civilian pilots just did not have a military pilot's flight time and experience. Something had to give. The minimum requirements for the airlines took a dive to meet the reality of what civilian pilots had to offer. The requirement for a four-year college degree was dropped. The requirement that pilots have 20/20 uncorrected vision was dropped. Minimum flight requirements plummeted.

The demand for pilots went up because more airlines went into operation. At the same time, the supply of pilots to meet the demand dropped because fewer military pilots were coming out. These market forces produced the "pilot boom." Airlines needed pilots more than pilots needed airlines; therefore, pilots could be selective.

Then the 1990s arrived. Tough economic times caused airlines to lose millions of dollars that forced many of them into bankruptcy. The reduced profitability of flights reduced the demand for pilots. The hiring requirements began to increase. Market forces shifted. Pilots needed airlines more than airlines needed pilots; therefore, the airlines could be selective.

I have had a unique view on all of this because I have trained many of the pilots who were the beneficiaries of the hiring boom. The shift in pilot requirements is clear. In 1987, I could not keep pilot graduates after they had 800 hours because they were hired almost automatically by commuter airlines. In 1988, former students needed only 1,000 hours. In 1989, it took about 1,200 hours, and the competition was tougher. By 1990, the target for marketability was 1,500–1,800 hours. In 1991, 1992, and 1993, more than 2,000 hours were required just to be considered.

The ratio of total time to multiengine time has also shifted. In 1987, not much more than a fresh multiengine rating was needed; the ratio was about 8-to-1 (1 hour of multiengine time for every 8 hours total). In 1990, the ratio was 5-to-1. Stated minimums as of this writing are approximately a 3-to-1 ratio.

Prospective pilots of the future must be keenly aware of the market fluctuations. The minimum flight times required for employment are constantly in transition and are driven by the economy, politics, and consumer demand.

BUILDING FLIGHT TIME

We have established that a large amount of total time and valuable multiengine time will be required to get an airline or corporate career off the ground. The most pressing question becomes "How do I get this flight time?" Pilots cannot simply buy 2,000 hours of flight time in both single-engine and multiengine airplanes. The expense would be astronomically high. Pilots must accumulate the time in other ways. Unfortunately, some of the "other ways" are illegal and getting caught could render your logbook and your aviation career dreams moot.

The need to build flight time actually starts after a pilot receives the private certificate. As a student pilot, the flight time is dictated by the flight instructor and the Federal Aviation Regulations. Private pilots know that they must gain flight time in order to get an instrument rating and a commercial certificate. Pilots need maximum flight hours on minimum dollars, and here some problems can arise. Let these next few pages be a guide to staying out of trouble and building acceptable flight time.

THE EXPENSE-SHARING MISUNDERSTANDING

The Federal Aviation Regulations (FARs) outline the privileges and limitations of a private pilot in FAR 61.118. Most private pilots have memorized a portion of FAR 61.118 that states: "A private pilot may share the operating expenses of a flight with his passengers." We are quick to realize that this regulation does not say that expenses must be shared equally. But there are some major misunderstandings about this rule.

A paragraph at the beginning of FAR 61.118 states: "A private pilot may not act as pilot in command of an aircraft that is carrying passengers or property for compensation or hire." The intent is to prevent private pilots from becoming paid professional pilots until they get further training and certificates. The "shared expenses" rule seems to be an exception to the "no compensation or hire" rule. When is it legal to share expenses with passengers and when is it not legal? The answer seems to lie within the original intent or reason for the flight. Look at these two similar examples.

Situation 1. A private pilot is walking out the door of the FBO to preflight a single-engine rented airplane. He is planning to fly to another airport 100 miles away in order to practice VFR navigation and to build his total and cross-country flight time. A friend of his sees him walking to the plane: "Do you mind if I ride along with you?" The private pilot remembers the expense-sharing rule and says, "Sure, you may ride along, but this flight is going to cost me about a hundred bucks. Do you mind chipping in a little of the cost?" The passenger agrees to help pay and they take off.

Situation 2: A relative of a private pilot calls and says that she must attend a meeting in a city 100 miles away. The relative asks if the private pilot could fly her to the city for the meeting. The private pilot says, "Sure, I'll fly you over there, but that trip will cost about a hundred bucks. Do you mind chipping in a little of the cost?" The relative agrees and the private pilot takes her to the meeting.

Are these flights legal? They could be to the same airport, they could cost the same, and they could have a passenger sharing expenses of the exact same dollar amount. Both situations sound the same. Both situations look the same to anyone watching the takeoff. But the FAA does not think they are the same. The FAA feels that the second situation is a direct violation of FAR 61.118.

How can pilots know how the FAA feels about a particular regulation? You might think that the simple answer is "Read the regulation." It is not that simple because different readers of the regulation might have a different interpretation of the regulation. Even FAA officials might give opposing interpretations on any specific regulation. The ultimate way to learn what the FAA regulations really mean is examination of what triggers the agency to prosecute a pilot.

In one FAA enforcement action against a private pilot, a neighbor called in the middle of the night reporting that a family member had fallen ill and wanted to be flown out immediately to visit. The private pilot and the neighbor took off at 4 a.m. on a mission of mercy. The private pilot asked the neighbor to share expenses of the flight, which the neighbor gladly did at the time. Later, the neighbor thought the shared cost of the flight was excessive. He called the FAA to ask what he thought was a harmless question about how much airplanes cost to fly. The FAA subsequently issued an emergency revocation of the private pilot's certificate.

During the investigation of this private pilot, the FAA turned up two other flights that they felt were violations. One took place when the private pilot flew in a guest speaker to his child's school. The private pilot took a tax deduction for the flight on the basis that the cost of the flight was his donation to a charity. The other flight in question involved the same private pilot transporting a lady to a distant city to consult with a physician about her medical treatment.

In the case of each flight, the FAA charged that the private pilot accepted compensation in the form of money or tax advantage that was outside the law. The FAA believed that the original intent of each flight was to transport another person. The only reason the flights were taken in the first place was to carry a passenger. If the respective passengers had not needed transportation, the flights would not have been made. The private pilot had not originally planned on making these flights, so shared expenses in these cases were illegal compensation.

The FAA issued a violation. A judge agreed with the FAA and suspended the private pilot's certificate. With airlines being very selective today, any violation on a pilot's FAA record means a professional pilot career can be forgotten.

"How can the FAA possibly know my true intent on any flight?" They can't. The FAA knows that people violate the shared-expense rule every day. That is why when they do catch somebody, they might "throw the book at them" in order to make that pilot an example to other violators who have not been caught yet. Each pilot must use good judgment with FAR 61.118 because your flight time investment and your hopes for a flying career rest on every flight.

Commercial considerations

Many pilots believe that when they become a commercial pilot, shared-expense issues can be forgotten. Nothing can be further from the truth. Commercial pilots have even more issues to resolve. FAR 61.139: "The holder of a commercial pilot certificate may act as pilot in command of an aircraft carrying passengers or property for compensation or hire." This regulation seems to leave the door open for just about anything. It also seems to say that commercial pilots can accept payments for their services as a pilot anytime a customer agrees to pay. What this regulation does *not* say is more important, and this can be a real pilot trap.

A portion of Part 135 regulations governs air-taxi operators, more commonly called charter pilots. To complete a flight that falls under FAR Part 135 rules, a pilot must have much more flight time and testing than it takes to get a commercial pilot certificate. If a commercial pilot ever makes a flight that could be considered a Part 135 operation without proper certification beyond the commercial certificate, the pilot is taking a tremendous risk.

What can a commercial pilot legally do? FAR Part 135.1 lists activities that require a commercial pilot certificate, but are not considered a Part 135 operation:

- Student instruction.
- Nonstop sightseeing flights that begin and end at the same airport and are conducted within a 25-statute mile radius of the airport.
- Ferry or training flights.

- Aerial work operations: crop dusting, seeding, spraying, and bird chasing; banner towing; aerial photography or survey; fire fighting; helicopter operations or repair work; power line or pipeline patrol.
- Sightseeing flights in hot air balloons.
- Nonstop flights conducted with a 25-statute mile radius of the airport of takeoff for the purpose of intentional parachute jumps.

There are other possibilities for helicopters, but this is the entire list for airplanes. Ideally, FAR 61.139 would have this Part 135 list under privileges and limitations of commercial pilots because this list better represents the legal privileges.

Situation 3. A commercial pilot flies a passenger to a basketball game at the passenger's request. The operating expense of the flight is $100. The commercial pilot charges the passenger $150. The commercial pilot pays the expense of the airplane and keeps $50 for services. Is this a legal situation?

No. When a person or property departs from one airport and lands at another airport on a flight that was solicited, a charter flight has occurred. This falls under Part 135 and a commercial pilot certificate alone does not qualify the pilot to make the flight.

Situation 4. A commercial pilot flies a passenger to a basketball game. The operating expense of the flight is $100. The commercial pilot charges the passenger $100. The commercial pilot pays the expense of the airplane, keeps no money for himself, and records the flight in a logbook. Is this a legal situation?

No. According to the FAA, and backed up in the courts, this also is a violation because the pilot did receive compensation. Rather than money, compensation was flight time in the logbook. Any compensation on a flight of this nature is illegal.

Situation 5. A commercial pilot flies a passenger to a basketball game. The operating expense of the flight is $100. The commercial pilot and passenger equally share the expense of the flight. Is this situation legal?

Part 61.139 on commercial pilot privileges does not make any mention of a commercial pilot sharing expenses. In practice, the FAA allows commercial pilots to share expenses as long as the "intent of the flight" issue is resolved properly. In Situation 5, if the only reason for the trip was to carry a passenger, then the flight was illegal.

Situation 6. A flight instructor flies an airplane carrying a passenger to a basketball game. The operating expense of the flight is $100. The flight instructor charges the passenger $100 for the airplane plus an instruction fee of $20 per hour for 2 hours and $5 for a new pilot logbook. The passenger pays a total of $145, and the instructor records the flight in the passenger's just-acquired logbook as dual instruction received. The instructor then records the time in his logbook under dual-instruction given. Is this a legal situation?

This one is tough. Clearly the instructor is using his instructor certificate to get around the Part 135 restriction. The passenger was not a student. Why did the instructor believe that this new student's first lesson needed to be a cross-country flight? This would be the FAA's first question. The FAA would surely contend that this is also a violation of Part 135, as well as an abuse of the flight instructor certificate. This pilot would be in deep trouble.

Situation 7. The commercial pilot/flight instructor is also a big basketball fan. Never mind.

An endless number of gray-area situations can be developed. The bottom line is not what the pilot thinks, but what the FAA will violate a pilot for; consider the FAA a sleeping giant. If you operate in these gray areas as a matter of practice and one day accidentally awaken the angry giant, he will not be in a good mood, and your certificates and career will hang by a thread.

Reality

What are the chances of getting caught? The FAA is understaffed and overworked. How can they possibly place every flight and every airport under surveillance? They can't. Chances are that if you are violating Part 135 regulations, the FAA won't catch you. Part 135 operators are self-policing.

Place yourself in the position of a Part 135 operator who has spent thousands of dollars on airplanes, certification, advertising, office supplies, telephone systems, and pilot training. You spend thousands of hours preparing manuals and pilot folders that are acceptable to the FAA for the certification. You cope with the FAA's surprise inspections and petty details. You are honestly trying to play by the rules.

One day you look out your office window and you see an airplane pull up to the FBO and unload passengers with brief cases. You will be suspicious, perhaps even calling the FAA to report the small airplane's tail number because this lowly commercial pilot might be stealing your business. If you are the pilot of the small airplane, beware. The chances of getting turned in are greater than the chances of the FAA outright catching you.

You might feel boxed in because apparently there aren't all that many legal ways to build flight time. Flight instruction is first on the commercial pilot privileges list (operations excluded from Part 135) and is probably the most readily available legal operation. There are geographic pockets where banner towing, aerial photography, or sightseeing flights are plentiful, but those pockets are not nationwide; many of these operations are also seasonal and cannot be counted on for long-term time building. The question remains: Where are pilots supposed to get valuable flight time?

I was interviewing a pilot once and the subject of flight time naturally came up in the conversation. She said that she had been working on building her time various ways. Her previous instructor told her, "If you do not have much flight time, you must create some." This very curious statement opens up all new gray areas of flight time. Without knowing anything was wrong, you might have fallen into some of the following traps.

FLIGHT SIMULATORS

Many questions come up about the use of flight simulators and how this time should be used. First, it is important to get terminology correct. According to FAA Advisory Circular 120-45A, an "*airplane simulator* is a full size replica of a specific type or make, model, and series airplane cockpit, including assemblages of equipment and programs necessary. The device must simulate the airplane in ground and flight operations, a force cuing system which provides cues at least equivalent to that of three degrees" of free motion. Most flight schools do not have anything that fits this description. Full-motion flight simulators are usually found in major airline training facilities or corporate aircraft training facilities. These simulators are boxed-in cockpits that stand high above the floor on legs. Hydraulics in the legs allow the entire box to pitch, roll, and yaw in response to the pilot's control inputs and simulated wind conditions.

The advisory circular offers one more definition. "An *airplane training device* is a full-scale replica of an airplane's instruments, equipment, panels, and controls in an open flight deck or an enclosed airplane cockpit, including assemblage of equipment and programs necessary to represent the airplane in ground and flight conditions to the extent of the systems installed in the device; does not require a force (motion) cuing or visual system." This definition describes training devices that do not move and are generic; they do not need to be an exact replica of a particular airplane.

Based on these definitions, most flight students use training devices, not flight simulators.

Can any simulator or training device time be used as flight time? No. FAR Part 1 defines flight time as, "the time from the moment the aircraft first moves under its own power for the purpose of flight until the moment it comes to rest at the next point of landing." Actual flight time can only be recorded if it takes place in an aircraft that moves. Part 1 defines an aircraft as "a device that is used or intended to be used for flight in the air." Flight simulators and training devices cannot move under their own power with the intention of flight; they cannot be used for flight in the air.

Yet there are still some instructors and even examiners who count simulator time as flight time. Where does the confusion come from? FAR 61.65 details the flight experience needed to qualify for an instrument rating. The regulation says that a pilot must have at least 125 hours of "pilot flight time." The applicant must also have had 40 hours of instrument instruction. Of these 40 hours, 20 hours can be in an approved instrument ground trainer (device). Also, 15 hours of the 40 must be "flight" instruction with an instrument flight instructor (CFII).

The applicant therefore must have 20 hours in an aircraft, and may have up to 20 hours in the ground trainer. The 20 flight hours must all be with an instructor: 15 hours with a CFII. None of these requirements reduces the applicant's total flight time requirement of 125 hours. A pilot is not eligible for an instrument rating with 105 flight hours and 20 hours of ground trainer time.

Several pilots have been stripped of the instrument rating because they got through an FAA checkride without the examiner seeing the problem. The FAA office in Oklahoma City never misses this mistake. Many more instrument applicants have been disappointed when a sharp examiner had to refuse to give the checkride under these conditions.

More confusion comes from FAR 61.129, dealing with the requirements for a commercial pilot certificate. A portion of FAR 61.129, titled "Flight Time as Pilot," requires "a total of at least 250 hours of flight time as pilot, which may include not more than 50 hours of instruction from an authorized instructor in a ground trainer." This all but suggests that, in this case, 50 hours of simulator time can be used as flight time, which leads to misinterpretation and mistakes in logbooks.

In practice, this regulation allows for 50 hours of simulator time in lieu of 50 hours flight time. The applicant can go to the commercial checkride with 200 hours of flight time and 50 hours ground trainer time. The instrument ground trainer requirements and the commercial ground trainer allowance are apparently contradictory. The only possible exception that would allow ground trainer time to count toward flight time is when it is specifically authorized by an official of the FAA, and only then under a strictly controlled and inspected environment.

How should flight simulator time be logged? Flight simulator and ground trainer time is logged under the column in most logbooks listing "training device" or "simulator." It also can be counted under "instruction" or "dual received" for the purpose of proving an applicant has had the allowed 20 hours toward an instrument rating or to prove that he is current for flight under IFR.

Simulator and ground trainer time can never be logged under "total duration of flight" or "flight time." In addition, all simulator and training device time must be signed off by a ground or flight instructor. FAR 61.189 requires the instructor to certify the time spent in a ground trainer by signing her name and giving her certificate number.

There is no such thing as "solo simulator." If you operate a simulator or ground trainer alone, the time spent cannot be used toward an instrument rating or to maintain instrument currency. If you have been logging simulator and ground trainer time as flight time, and that time is included in what you are calling "total time," you should subtract the time.

If you land an interview with a company and they subsequently find out that you logged time inappropriately, the hiring process will conclude very quickly. If you have used simulator or ground trainer time to qualify for an FAA certificate that specifically requires flight time, your certificate is in jeopardy, even if you have the proper flight time now. Double-check your logbooks. Do not let an airline interviewer or flight examiner spot this problem before you do.

SAFETY PILOTS

Some pilots have built flight time by creating "safety pilot" time. (If you want to start a fight, just walk into any hangar in America and take either side of this controversy.) In order for instrument-rated pilots to fly in IFR conditions, they must have logged 6 hours of actual or simulated instrument time (FAR 61.57). Of the 6 hours, 3 hours can be in a ground trainer. Also, at least 6 instrument approaches are required, and they can all be in the simulator. This must be accomplished within any 6-month time block.

If a pilot fails to get this experience within the 6-month period, he can no longer act as pilot in command under IFR. The FAA then gives the pilot an additional 6-month grace period to get current. During the grace period, if the pilot lacks the portion of the currency requirement that must be accomplished in an airplane, instrument conditions must be simulated while flying in VFR conditions.

It is not safe to fly solo with an IFR hood on; someone should be onboard to watch for traffic and other potential problems. This is where the safety pilot comes in. FAR 61.51 says, "A pilot may log as instrument flight time only that time during which he operates the aircraft solely by reference to instruments, under actual or simulated instrument flight conditions. Each entry must include the place and type of each instrument approach completed, and the name of the safety pilot for each simulated instrument flight." A safety pilot is someone who will look for traffic while the instrument pilot flies under the hood. A safety pilot must be rated and current in the airplane used.

The pilot flying under the hood may log pilot-in-command time for this flight, but can the safety pilot log any flight time? FAR 61.51 says, "A pilot may log as second-in-command time all flight time during which he acts as second-in-command of an aircraft on which more than one pilot is required under the type certification of the aircraft, or the regulations under which the flight is conducted." Under FAR 61.51, a safety pilot is required.

So, when conducting a flight under that regulation, two pilots are required, and 61.51 says that any time two pilots are required, one pilot can log second in command. Second-in-command time does increase the total flight time. This means that two private pilots flying a Cessna 150 could both log flight time as long as one pilot flies under the hood.

Apply this idea to the concept of quality flight time. I do not feel the Cessna 150 second-in-command time is very valuable. If a potential employer calls you in for an interview, the employer will be looking for the quality of your time, not necessarily your quantity. If you have just barely met the airline's or Part 135 operator's minimum time requirements on the strength of single-engine second-in-command time, your application will be disregarded. Put only quality pilot-in-command time in your logbook.

Can a safety pilot log the time as pilot in command? This is up for interpretation. It is possible for two pilots in the same airplane to both log pilot in command for simultaneous flight time. But FAR 61.51 allows it only under the condition where "a certificated flight instructor may log as pilot-in-command time all flight time during which he acts as a flight instructor." When a certified instructor is instructing in flight, he logs the time PIC, and if the pilot receiving instruction is at least a private pilot in an aircraft that she is rated to fly, she also can log PIC. By this regulation, it appears that the only time two pilots can simultaneously log PIC is when one is an instructor. If the safety pilot is not an instructor, then it seems that pilot-in-command time would not be allowed.

If you are planning a flight career or just maintaining a pristine logbook, it is best to stay away from any questionable flight time situations. You might feel in your heart that the logged flight time is valid, but a heart has nothing to with it. The only thing that counts is what an airline interviewer or flight examiner can accept as fact and use that fact to hire you or issue a fresh rating or certificate.

Controversial flight time is like poison ivy to the airlines and air-taxi operators. They have enough problems without dealing with the FAA in some future investigation about your flight time. They will always take a pilot with a spotlessly clean logbook over one that has built time in the gray areas. Do not be short-sighted and grab for the quick time over the quality time.

HITCH-HIKING ON 135

Many pilots have built time while riding along on Part 135 air-taxi airplanes. This also creates problems. First, if you do not meet the Part 135 air-taxi flight time minimums, you cannot log anything as PIC that is truly a Part 135 operation. FAR Part 135.243 requires a pilot to have 500 hours of pilot time to qualify as a VFR-only Part 135 pilot in command. The same regulation requires 1,200 hours for IFR pilot-in-command operations. Even if you meet these minimums, you still cannot log the flight time unless you actually work for and have been tested by the Part 135 operator.

Take a situation where a charter airplane leaves airport A and flies to airport B. At airport B, paying passengers are picked up and flown to airport C. A legal and certified IFR pilot is flying the airplane. The Part 135 pilot's best friend, who is also a pilot, but not employed by the air-taxi company, is riding along. Is the friend who rides along able to log any of this trip? Yes, but only that portion of the flight that could be considered under Part 91 flight rules. The ride-along friend cannot log portions of the flight conducted under Part 135 flight rules.

The leg from airport A to airport B was essentially a deadhead leg. No passengers or goods were on board, so there was no requirement for Part 135. The pilots could agree before takeoff that the ride-along pilot is PIC for the first leg. The leg from airport B to airport C with passengers aboard does require a qualified Part 135 PIC; the ride-along pilot does not qualify and cannot log the flight time.

Then deadhead legs on Part 135 airplanes are not technically Part 135 operations and are outside the FAA's Part 135 jurisdiction. Deadhead ride-along pilots might be prohibited by the Part 135 operator's procedures manual or prohibited by an insurance carrier. A Part 135 pilot might be placing his job and career on the line by allowing pilot friends to ride along, even though it appears to be a big favor for the friends.

A ride-along pilot needs to know that the Part 135 operator will not pay higher insurance premiums to protect an unneeded pilot. If anything should happen on the flight, a ride-along pilot will not be covered under any insurance. If an accident were to occur, the Part 135 operator could even take a ride-along pilot to court and claim that his inexperience caused the accident. The Part 135 operator could claim that the ride-along was an illegal stowaway. The case might be in court for years and legal fees for a ride-along might mount unchecked.

It's not all bad. Selected Part 135 operators will approve and even encourage ride-along pilots. They use riding along as an inexpensive way to train future pilots. Stay above board. Ask the Part 135 operator about the company policy on rider pilots. For your own protection, ask to inspect the company's current Part 135 operating manuals and the company's current insurance policy. If you spot a deficiency that puts you at risk in any way, or if the operator refuses access, do not fly with that air-taxi. If logging deadhead legs is approved by the operator, and you are completely satisfied that the air-taxi is legal, safe, and properly insured, the exposure might help you find a job there in the future.

MILITARY CROSSOVER TIME

If you are a military pilot and seek a job in the civilian flight world, you will need to transfer your military flight experience to a civilian pilot certificate. The process involves taking your military flight records to a local FAA flight standards district office and having the records qualified. The FAA inspector will give a commercial pilot written exam, and after passing the exam, the military pilot is awarded the commercial pilot certificate. The process is straightforward, but the FAA is very stingy with anything that they do not consider flight time.

A controversy has come up over military flight officers who fly in the back seat of a military airplane as weapons systems officers (WSO). These people are trained to navigate all over the world, monitor airplane systems, and operate the armaments. But the FAA does not consider WSOs pilots because they do not actually fly the airplane. Certain backseats do have sticks so that WSOs can fly the airplane, and some officers do, but it is not recognized by the FAA.

A WSO was upset that his time was not considered flight time by the local FAA. He started calling around to other FSDO offices and finally found an FAA inspector who could be convinced that WSO time was pilot time. The WSO cited FAR 61.155 pertaining to flight hours used to qualify as an airline transport pilot. FAR 61.155 states, in part: "A commercial pilot may credit the following flight time toward the 1,500 hours total flight time requirement . . . with flight engineer time acquired in airplanes required to have a flight engineer by their approved aircraft flight manuals, while participating at the same time in an approved pilot training program."

FAR 61.155 continues: "The applicant (for the ATP) may not credit more than one hour for each three hours of flight engineer flight time so acquired, nor more than a total of 500 hours." The WSO argued that backseat fighter time was even more valuable than flight engineer time because backseaters do have a control stick; therefore, backseaters should have at least the same 1-to-3 flight hour ratio as flight engineers. The FAA inspector accepted the argument.

The WSO drove 400 miles the next day to that inspector's office. His WSO-backseat time was the basis for flight time credit toward the airline transport pilot certificate. The WSO drove home the day after, and the word spread of his success. Soon, a steady stream of WSOs was quietly driving the 400 miles to see this one inspector. It did not last long.

The inspector finally stopped issuing the logbook certifications. The first WSO is an employed pilot. All other WSOs must start as student pilots with zero flight time when they become civilians.

FALSE RECORDS

Of course there are pilots who have more flight time in their logbooks than they have actually flown. FAR 61.59 speaks specifically: "No person may make or cause to be made any fraudulent or intentionally false entry in any logbook, record, or report that is required to be kept, made, or used, to show compliance with any requirement for the issuance, or exercise of the privileges, or any certificate or rating under this part."

If a pilot is caught, according to 61.59: "The commission by any person of an act prohibited under (the paragraph regarding logbook falsification) of this section is a basis for suspending or revoking any airman or ground instructor certificate or rating held by that person." Many pilots believe that the chances of getting caught are so small that there is no risk involved by padding the logbook. The FAA knows that pilots cheat in the logbooks every day. The agency also knows that they cannot catch everyone. Pilots that do get caught face maximum enforcement by the agency.

Several flight instructors working together at an FBO lost their certificates due to falsifying records. The instructors were accused of adding flight time to individual flights. If a lesson took 1.2 hours, they might put 1.5 hours in their logbook. They were also accused of logging the wrong airplanes. A lesson might have been accomplished in a Sundowner, but they listed the flight in a Bonanza.

The FAA proved that the flight instructors falsified their logbooks by obtaining a subpoena for the pilots' logbooks, airplanes' logbooks, and FBO invoices. It became a matter of mathematics. The total time claimed by the instructors during one period of time was greater than the total hours on the school's airplanes during the same time period.

The FAA stripped the instructors of all their pilot certificates and ratings and banned them from flying for one year. This meant they would have to wait a full year to start all over again as student pilots. Don't spoil your future with illegal logbook entries.

LOGBOOK CERTIFICATION

Every would-be professional pilot will turn over their logbooks to the FAA at least once. The action is required for an ATP certificate. After accumulating 1,500 hours, pilots schedule an appointment with an FAA inspector. The pilot must bring the logbook and a valid first-class medical certificate to the meeting.

The inspector will go through each page of the logbook and ask questions about many entries. If it is approved, the inspector will certify the logbook and issue a permit to take the airline transport pilot written test one time. You do not want to fail this test. If you do fail, you must go back to the FAA and get another test permit, and answer the inevitable questions about why you failed.

Getting your logbook FAA-certified is a good idea. If a potential employer has a question about any of your flight time, the certification shows acceptability by the FAA.

Logbooks are also submitted for Part 135 review. In one case, a pilot who had been riding along on previous Part 135 flights had made 22 PIC entries. The FAA believed this to be suspicious because of the amount of time recorded and took action against the pilot. The pilot testified at a hearing that he had only flown the airplane during the deadhead legs, the technique that was considered earlier in this chapter.

He logged the entire flight as PIC, including the portions that were Part 135 operations. A judge ruled that the logbook was intentionally false and ordered that his commercial pilot certificate be revoked. The case was appealed to the National Transportation Safety Board, but the board upheld the judge's order by saying that the pilot had "misrepresented" his flight experience.

Pilots are always weighing risk in their piloting decisions. The risk of getting caught falsifying records might be small, but for future professional pilots, the risk is greater than most pilots think. The possible advantage a pilot gets by falsifying the logbook is just not worth gambling an entire career over.

The safest path to unquestionable flight time is by strict adherence to the regulations, especially recognizing the importance of FAR 135.1. This usually means a couple of years working as a flight instructor before working for a Part 135 operator. Multiengine time will mount with the Part 135 operator.

CAREER CLIMBING

The most likely career ladder starts with the commercial certificate, then a flight instructor certificate (and flight instruction) to log the quantity of flight time, then airtaxi work as a Part 135 pilot to get the quality multiengine time. So, the key to the best marketability seems to go through charter flying.

Unfortunately charter and air-taxi is very sensitive to the business cycle. When business and corporations experience hard times, one of the first things cut out is travel. This means fewer charter flights and fewer multiengine hours for up-and-coming pilots. Part 135 flying is a rung on the pilot career ladder that can become a roadblock.

Corporate aviation is another way to build flying time. Business flying might pan out as a full-time career. The problem is that many of these jobs require a good deal of flight time, perhaps an ATP certificate and a type rating, to meet insurance and FAA requirements. Even so, many pilots have been at the right place at the right time to land a corporate job that provided accelerated flight time.

The best way to accumulate a large quantity of single-engine time and enough quality multiengine time to become job marketable is to work hard, and stay out of the gray areas.

Index

Illustration page numbers are in boldface type.